普通高等学校规划教材

机电工程英语实用教程

主　编：邱　晋　廖巧云
副主编：王　敏　杨　俊
主　审：毛明勇

人民交通出版社股份有限公司
北　京

内 容 提 要

本书共分为四大部分:第一部分为开篇引介,涉及电气工程和机械工程发展历史、适用领域、功能用途等;第二部分为电气工程知识,涉及基础的电学、电气知识,电力系统发展,电磁现象及与之相关的电机、变压器等,并介绍了电气工程在智能家居、电动汽车及量子悬浮等领域的运用;第三部分为机械工程知识,涉及基本的机制及机械零件,工程材料及其性能介绍,各种金属加工工艺等,并介绍了机械工程在工业机器人、纳米技术及3D打印等领域的运用;第四部分为前景展望,分别对电气工程和机械工程领域的发展前景及其对人类的影响等方面进行了前瞻性展望。本书在每一小节后都配有不同类型的相关练习,内容涵盖篇章理解、词汇巩固和翻译强化,并附有参考答案。

本教材实用性较强,既适合高等院校专业英语教学,又适合具备一定英语水平者自学。

图书在版编目(CIP)数据

机电工程英语实用教程 / 邱晋,廖巧云主编.—北京:人民交通出版社股份有限公司,2022.7
 ISBN 978-7-114-17913-6

Ⅰ.①机… Ⅱ.①邱… ②廖… Ⅲ.①机电工程—英语—高等学校—教材 Ⅳ.①TH

中国版本图书馆 CIP 数据核字(2022)第 062666 号

Jidian Gongcheng Yingyu Shiyong Jiaocheng

书　　名:	机电工程英语实用教程
著 作 者:	邱　晋　廖巧云
责任编辑:	钱悦良
责任校对:	孙国靖　宋佳时
责任印制:	刘高彤
出版发行:	人民交通出版社股份有限公司
地　　址:	(100011)北京市朝阳区安定门外外馆斜街3号
网　　址:	http://www.ccpcl.com.cn
销售电话:	(010)59757973
总 经 销:	人民交通出版社股份有限公司发行部
经　　销:	各地新华书店
印　　刷:	北京交通印务有限公司
开　　本:	787×1092　1/16
印　　张:	18.25
字　　数:	400千
版　　次:	2022年7月　第1版
印　　次:	2022年7月　第1次印刷
书　　号:	ISBN 978-7-114-17913-6
定　　价:	45.00元

(有印刷、装订质量问题的图书由本公司负责调换)

前　言

在长期的机电工程英语教学过程中,我们发现市面上已出版的各类机电工程方向的双语或英语教材虽然数量不少,但有的难度较大,课时要求过多;有的教材内容中英语表达过于简单,不适用于英语专业学生的学习;还有的编写时间较早,部分知识点亟待更新。有鉴于此,我们教学团队经过几年的构思,群策群力编写出这本适用于英语专业学生的《机电工程英语实用教程》。

本书的编写融入了当前机电工程方向的最新发展趋势,编排体例循序渐进,在用英语普及电气工程和机械工程基础知识的同时,还系统地从词汇、句法到语篇全方位地介绍了科技英语的翻译技巧和方法。另外,本书在每一小节后都配有不同类型的相关练习,内容涵盖篇章理解、词汇巩固和翻译强化,并附有参考答案。因此,对于具备基础英语知识的学习者来说,这是一本既适合专业英语教学又适合自学提升英语应用能力的实用性教材。

本书共分为四大部分:

第一部分开篇引介为第一章。本部分涉及电气工程和机械工程发展历史、适用领域、功能用途等内容。

第二部分为电气工程知识,包括第二章到第五章。本部分涉及基础的电学、电气知识,电力系统发展,电磁现象及与之相关的电机、变压器等,并介绍了电气工程在智能家居、电动汽车及量子悬浮等领域的运用。

第三部分为机械工程知识,包括第六章到第九章。本部分涉及基本的机制及机械零件,工程材料及其性能介绍,各种金属加工工艺等,并介绍了机械工程在工业机器人、纳米技术及3D打印等领域的运用。

第四部分前景展望为第十章。本部分分别对电气工程和机械工程领域的发展前景及其对人类的影响等方面进行了前瞻性展望。

本书在编写过程中参考引用了国内外大量专家学者的教材、专著、文章等。

虽然尽力注明,但由于篇幅限制,部分文献在书中并未得到一一注明,在此致以诚挚的谢意和歉意。另外,由于编者水平有限,书中疏误在所难免,恳请专家同行和广大读者批评指正。

本教材是重庆交通大学规划教材立项建设项目的成果。主编为重庆交通大学外国语学院邱晋副教授、上海外国语大学语言研究院廖巧云教授,副主编为重庆交通大学王敏副教授、杨俊副教授,主审为重庆交通大学毛明勇教授。本书在编写期间,得到重庆交通大学各部门以及人民交通出版社股份有限公司各位编辑的热心帮助和支持,在此表示衷心感谢。

编　者

2021 年 12 月

CONTENTS

PART I Introduction ·· (1)

  Unit 1　Introduction to Engineering ··· (3)

    Before-Class Reading ·· (3)

    Section A　What Is Engineering? ··· (3)

    In-Class Reading ··· (7)

    Section B　Introduction to Electrical Engineering ······························· (7)

    Section C　Introduction to Mechanical Engineering ·························· (14)

    Translation Skill I ··· (21)

    科技英语翻译学习的理论基础——以系统功能语言学为例 ············· (21)

PART II Electrical Engineering ·· (23)

  Unit 2　Electricity ··· (25)

    Before-Class Reading ·· (25)

    Section A　Electricity in Early Days ·· (25)

    In-Class Reading ··· (29)

    Section B　Electrons and Electricity ·· (29)

    Section C　Electricity and Electronics ··· (37)

    Translation Skill II ·· (44)

    科技英语翻译的原则和标准 ··· (44)

  Unit 3　Power Source ·· (46)

    Before-Class Reading ·· (46)

    Section A　Direct Current and Alternating Current ························· (46)

    In-Class Reading ··· (49)

    Section B　Electrical Power System ·· (49)

    Section C　Hydropower Technology Development ··························· (58)

Translation Skill Ⅲ ……………………………………………………………… (64)

科技英语的特点 ……………………………………………………………… (64)

Unit 4　Electromagnetism …………………………………………………… (68)

Before-Class Reading ………………………………………………………… (68)

Section A　Introduction to Electromagnetism …………………………… (68)

In-Class Reading ……………………………………………………………… (74)

Section B　Direct Current Generators and Motors ……………………… (74)

Section C　Transformers …………………………………………………… (81)

Translation Skill Ⅳ …………………………………………………………… (85)

词语的处理(1)——词义的引申 …………………………………………… (85)

Unit 5　Electrical Engineering Applications ………………………………… (88)

Before-Class Reading ………………………………………………………… (88)

Section A　Exploring Home Automation and Domotics ………………… (88)

In-Class Reading ……………………………………………………………… (93)

Section B　Everything Explained about Electric Cars: Basics of an EV …… (93)

Section C　How Does Quantum Levitation Work? ……………………… (100)

Translation Skill Ⅴ …………………………………………………………… (105)

词语的处理(2)——词性的转换 …………………………………………… (105)

PART Ⅲ　Mechanical Engineering ………………………………………… (109)

Unit 6　Basic Mechanisms and Elements …………………………………… (111)

Before-Class Reading ………………………………………………………… (111)

Section A　Brief Introduction to Mechanism …………………………… (111)

In-Class Reading ……………………………………………………………… (113)

Section B　Transmission …………………………………………………… (113)

Section C　Basic Machine Elements ……………………………………… (121)

Section D　Joints and Linkages …………………………………………… (132)

Translation Skill Ⅵ …………………………………………………………… (137)

词语的处理(3)——增词与省略 …………………………………………… (137)

Unit 7　Engineering Materials ………………………………………………… (140)

Before-Class Reading ………………………………………………………… (140)

Section A　General Properties of Engineering Materials ………………… (140)

In-Class Reading ……………………………………………………………… (149)

Section B　Metallic Materials ……………………………………………… (149)
Section C　Nonmetallic Materials …………………………………………… (157)
Translation Skill Ⅶ …………………………………………………………… (163)
句法的处理(1)——被动语态的翻译 ………………………………………… (163)

Unit 8　Metal Working Processes ……………………………………………… (166)
Before-Class Reading ………………………………………………………… (166)
Section A　Heat Treatment …………………………………………………… (166)
In-Class Reading ……………………………………………………………… (171)
Section B　Forming …………………………………………………………… (171)
Section C　Casting …………………………………………………………… (179)
Translation Skill Ⅷ …………………………………………………………… (187)
句法的处理(2)——否定句的翻译 ………………………………………… (187)

Unit 9　Mechanical Engineering Applications ……………………………… (189)
Before-Class Reading ………………………………………………………… (189)
Section A　Most Popular Industrial Robotic Applications for 2021
　　　　　 and Projections ………………………………………………… (189)
In-Class Reading ……………………………………………………………… (195)
Section B　What Is Nanotechnology? ……………………………………… (195)
Section C　3D Printing：History，Processes and Future ……………………… (201)
Translation Skill Ⅸ …………………………………………………………… (207)
句法的处理(3)——从句的翻译 …………………………………………… (207)

PART Ⅳ　Prospect ………………………………………………………………… (211)

Unit 10　Prospect of Electrical and Mechanical Engineering ……………… (213)
Before-Class Reading ………………………………………………………… (213)
Section A　Electrical and Electronic Engineering in the Future …………… (213)
In-Class Reading ……………………………………………………………… (219)
Section B　The Future of Mechanical Engineering：A Vision and a Mission ……… (219)
Translation Skill Ⅹ …………………………………………………………… (224)
科技英语语篇的翻译 ………………………………………………………… (224)

Glossary ……………………………………………………………………… (227)

参考文献 ……………………………………………………………………… (251)

参考答案 ……………………………………………………………………… (252)

PART I
Introduction

Unit 1

Introduction to Engineering

Before-Class Reading

Answer the following questions before reading the passage:
1. Do you know the names of your university, your department, and your orientation?
2. Are there other colleges of engineering in your school? What are they?

Section A
What Is Engineering?

The word "engineering" derives from the Latin root *ingeniere*, meaning to design or to devise, which also forms the basis of the word "ingenious". Those meanings are quite appropriate summaries of the traits of a good engineer. At the most fundamental level, engineers apply their knowledge of mathematics, science, and materials—as well as their skills in communications and business—to develop new and better technologies. Rather than experiment solely through trial and error, engineers are educated to use mathematics, scientific principles, and computer simulations as tools to create faster, more accurate, and more economical designs.

In that sense, the work of an engineer differs from that of a scientist, who would normally emphasize the discovery of physical laws rather than apply those phenomena to develop new products. Engineering is essentially a bridge between scientific discovery and product applications. Engineering does not exist for the sake of furthering or applying mathematics, science, and computation by themselves. Rather, engineering is a driver of social and economic growth and an integral part of the business cycle. With that perspective, the U.S. Department of Labor summarizes the engineering profession as follows:

Engineers apply the theories and principles of science and mathematics to research and develop economical solutions to technical problems. Their work is the link between perceived social needs and commercial applications. Engineers design products, machinery to build those products, plants in which those products are made, and the systems that ensure the quality of the products and the efficiency of the workforce and manufacturing process. Engineers design, plan, and supervise the construction of buildings, highways, and transit systems. They develop and implement improved ways to extract, process, and use raw materials, such as petroleum and natural gas. They develop new materials that both improve the performance of products and take advantage of advances in technology. They harness the power of the sun, the earth, atoms, and electricity for use in supplying the Nation's power needs, and create millions of products using power. They analyze the impact of the products they develop or the systems they design on the environment and on people using them. Engineering knowledge is applied to improving many things, including the quality of healthcare, the safety of food products, and the operation of financial systems.

Historically, the main branches of engineering are categorized as follows:

Chemical engineering—It is the application of physics, chemistry, biology, and engineering principles in order to carry out chemical processes on a commercial scale.

Civil engineering—It is the design and construction of public and private works, such as infrastructure, bridges, tunnels, dams, and buildings.

Electrical engineering—It is the design, study, and manufacture of various electrical and electronic systems.

Mechanical engineering—It is the design and manufacture of physical or mechanical systems.

With the rapid advancement of technology, many new fields are gaining prominence and new branches are developing, such as computer engineering, software engineering, nanotechnology, molecular engineering, mechatronics, etc. These new specialties sometimes combine with the traditional fields and form new branches, such as mechanical engineering and mechatronics, electrical and computer engineering.

(513 words)
From *An Introduction to Mechanical Engineering*
by Wickert, J. & Lewis, K.

New Words and Expressions

atom ['ætəm] n.	原子
harness ['hɑːnis] v.	利用；治理
infrastructure ['infrəstrʌktʃə] n.	基础设施；下部构造
ingenious [in'dʒiːniəs] a.	精巧的，巧妙的，新颖独特的
integral ['intigrəl] a.	完整的，不可或缺的
mechatronics [ˌmekə'trɔniks] n.	机械电子学，机电一体化
nanotechnology [ˌnænəutek'nɔlədʒi] n.	纳米技术
performance n.	性能
prominence ['prɔminəns] n.	显著，突出，卓越
simulation [ˌsimju'leiʃn] n.	模拟，仿真
molecular [mə'lekjələ] engineering	分子工程学
raw material	原材料
transit ['trænzit] system	公交系统；转接系统
trial and error	试错法

Exercises

📖 Text Understanding

I. Decide whether the following statements are true (T) or false (F) according to the passage.

1. The word "engineering" means to design or to devise.
2. All the fundamental skills for engineers are mathematical skill, scientific skill and material skill.
3. An engineer's work is almost the same as a scientist's work.
4. Trial and error experiment is of practical use in engineering.
5. Mechatronics is a part of mechanical engineering.

II. Fill in the blanks with proper words according to the passage.

1. Engineering is essentially a bridge between _____ and _____.
2. Engineers' work is the link between _____ and _____.
3. Engineering knowledge is applied to improving many things, including _____,

_____, and _____.

4. Historically, the main branches of engineering are _____, _____, _____, and _____.

5. _____ is the design, study, and manufacture of various electrical and electronic systems.

III. Give brief answers to the following questions.

1. What is the practical function of engineering?

_____

2. Give specific examples to illustrate the work done by civil engineers.

_____

Vocabulary Building

IV. Fill in the table below by giving the corresponding translation.

English	Chinese
molecular engineering	
transit system	
	原子
	纳米技术
trial and error	
	机械电子学
	模拟,仿真

V. Fill in the blanks with the words from the passage. The first letter of the word is given.

1. An appropriate summary of the traits of a good engineer can be i_____.
2. Engineering is considered as an i_____ part of the business cycle.
3. Their work is the l_____ between perceived social needs and commercial applications.
4. Engineers develop new materials that both improve the p_____ of products and take advantage of advances in technology.
5. With the rapid advancement of technology, many new fields are gaining p_____ and new branches are developing.

In-Class Reading

Section B
Introduction to Electrical Engineering

Electrical engineering is an engineering discipline concerned with the study, design, and application of equipment, devices, and systems which use electricity, electronics, and electromagnetism. It emerged as an identifiable occupation in the latter half of the 19th century after commercialization of the electric telegraph, the telephone, and electrical power generation, transmission, distribution, and use.

Subfields of Electrical Engineering

Electrical engineering has many subdisciplines, the most common of which are listed below:

Power engineering   Power engineering deals with the generation, transmission, and distribution of electricity as well as the design of a range of related devices. These include transformers, electric generators, electric motors, high voltage engineering, and power electronics. In many regions of the world, governments maintain an electrical network called a power grid that connects a variety of generators together with users of their energy. Users purchase electrical energy from the grid, avoiding the costly exercise of having to generate their own. Power engineers may work on the design and maintenance of the power grid as well as the power systems that connect to it. Such systems are called on-grid power systems and may supply the grid with additional power, draw power from the grid, or do both. Power engineers may also work on systems that do not connect to the grid, called off-grid power systems, which in some cases are preferable to on-grid systems. The future includes satellite controlled power systems, with feedback in real time to prevent power surges and prevent blackouts.

Control engineering   Control engineering focuses on the modeling of a diverse range of dynamic systems and the design of controllers that will cause these systems to behave in the desired manner. To implement such controllers, electrical engineers may use electronic circuits, digital signal processors, microcontrollers, and programmable logic controllers (PLCs). Control engineering has a wide range of applications from the flight and propulsion systems of commercial airliners to the cruise control present in many modern automobiles. It also plays an important role in industrial automation.

Control engineers often utilize feedback when designing control systems. For example, in an automobile with cruise control, the vehicle's speed is continuously monitored and fed back to the system which adjusts the motor's power output accordingly. Where there is regular feedback, control theory can be used to determine how the system responds to such feedback.

Electronic engineering　　Electronic engineering involves the design and testing of electronic circuits that use the properties of components such as resistors, capacitors, inductors, diodes, and transistors to achieve a particular functionality. The tuned circuit, which allows the user of a radio to filter out all but a single station, is just one example of such a circuit. Another example to research is a pneumatic signal conditioner.

Prior to the Second World War, the subject was commonly known as radio engineering and basically was restricted to aspects of communications and radar, commercial radio, and early television. Later, in post-war years, as consumer devices began to be developed, the field grew to include modern television, audio systems, computers, and microprocessors. In the mid-to-late 1950s, the term radio engineering gradually gave way to the name electronic engineering.

Before the invention of the integrated circuit in 1959, electronic circuits were constructed from discrete components that could be manipulated by humans. Although these discrete circuits consumed much space and power and were limited in speed, they are still common in some applications. By contrast, integrated circuits packed a large number—often millions—of tiny electrical components, mainly transistors, into a small chip around the size of a coin. This allowed for the powerful computers and other electronic devices we see today.

Telecommunications engineering　　Telecommunications engineering focuses on the transmission of information across a communication channel such as a coaxial cable, optical fiber, or free space. Transmissions across free space require information to be encoded in a carrier signal to shift the information to a carrier frequency suitable for transmission; this is known as modulation. Popular analog modulation techniques include amplitude modulation and frequency modulation. The choice of modulation affects the cost and performance of a system and these two factors must be balanced carefully by the engineer.

Once the transmission characteristics of a system are determined, telecommunication engineers design the transmitters and receivers needed for such systems. These two are sometimes combined to form a two-way communication device known as a transceiver. A key consideration in the design of transmitters is their power consumption as this is closely related to their signal strength. Typically, if the power of the transmitted signal is insufficient once the signal arrives at the receiver's antenna(s), the information contained in the signal will be corrupted by noise.

Instrumentation engineering　　Instrumentation engineering deals with the design of devices to measure physical quantities such as pressure, flow, and temperature. The design of such instruments requires a good understanding of physics that often extends beyond electromagnetic theory. For example, flight instruments measure variables such as wind speed and altitude to enable pilots the control of aircraft analytically. Similarly, thermocouples use the Peltier-Seebeck effect to measure the temperature difference between two points.

Often instrumentation is not used by itself, but instead as the sensors of larger electrical systems. For example, a thermocouple might be used to help ensure a furnace's temperature remains constant. For this reason, instrumentation engineering is often viewed as the counterpart of control.

Present Work of Electrical Engineers

From the Global Positioning System (GPS) to electric power generation, electrical engineers have contributed to the development of a wide range of technologies. They design, develop, test, and supervise the deployment of electrical systems and electronic devices. For example, they may work on the design of telecommunication systems, the operation of electric power stations, the lighting and wiring of buildings, the design of household appliances, or the electrical control of industrial machinery.

However, for many engineers, technical work accounts for only a fraction of the work they do. A lot of time may also be spent on tasks such as discussing proposals with clients, preparing budgets, and determining project schedules. Many senior engineers manage a team of technicians or other engineers, and for this reason project management skills are also important. As a result, most engineering projects today involve some form of documentation and strong written communication skills.

(1043 words)

https://encyclopedia.thefreedictionary.com/electrical+engineering

New Words and Expressions

altitude ['æltitjuːd] n.	海拔,高度
analog ['ænəˌlɔg] a. & n.	模拟的 & 模拟;类似物
antenna [æn'tenə] n.	天线
blackout n.	断电;灯火熄灭
capacitor [kə'pæsitə] n.	电容器
carrier n.	载波;载体
component [kəm'pəunənt] n.	组件,元件
counterpart n.	对应的事物,配对物

deployment [diˈplɔimənt] n.	部署，调度
diode [ˈdaiəud] n.	（电子）二极管
distribution n.	分配，供应；分布
dynamic [daiˈnæmik] a.	动态的，动力的
electromagnetic [iˌlektrəumægˈnetik] a.	电磁的
electromagnetism [iˌlektrəuˈmægnətizəm] n.	电磁，电磁学
electronics [iˌlekˈtrɔniks] n.	电子学
generate [ˈdʒenəreit] v.	发电；产生，生成
generation [ˌdʒenəˈreiʃn] n.	发电
generator [ˈdʒenəreitə] n.	发电机
grid [grid] n.	输电网；网格
inductor [inˈdʌktə] n.	电感器，感应器
maintenance [ˈmeintənəns] n.	维护，保养；保持
modulation [ˌmɔdjuˈleiʃn] n.	调制，调谐
motor [ˈməutə] n.	电机，电动机，马达
pneumatic [njuːˈmætik] a.	气动的，风动的
propulsion [prəˈpʌlʃn] n.	推动力，推进
resistor [riˈzistə] n.	电阻器
sensor [ˈsensə] n.	传感器，敏感元件
thermocouple n.	热电偶
transceiver [trænˈsiːvə] n.	无线电收发两用机
transformer [trænsˈfɔːmə] n.	变压器
transistor [trænˈsistə] n.	晶体管
transmission [trænsˈmiʃn] n.	传送，传递；变速器
transmitter [trænzˈmitə] n.	发送器；发射机
variable [ˈveəriəbl] n. & a.	变量 & 可变的
amplitude modulation	振幅调制
coaxial [kəuˈæksiəl] cable	同轴电缆
control engineering	控制工程
cruise control	定速巡航，巡航控制
discrete [diˈskriːt] component	离散元件，分立元件
electric circuit	电路
electrical power	电力；电源；电功率
electronic circuit	电子线路
electronic engineering	电子工程
give way to	为……所替代，让位于

high voltage	高压
household appliance	家用电器,家电
instrumentation engineering	仪表工程
integrated circuit	集成电路
optical fiber	光纤,光导纤维
power engineering	电力工程;动力工程
power surge	电力高峰
power system	电力系统
telecommunications engineering	通信工程,电信工程
tuned circuit	调谐电路

Notes

1. Global Positioning System (GPS)　　全球定位系统
2. Peltier-Seebeck effect　　珀尔帖—塞贝克效应
3. programmable logic controllers (PLCs)　　可编程逻辑控制器

📖 Text Understanding

Ⅰ. Choose the best answer according to the passage.

1. A(n) _____ connects a variety of generators with energy users.
   A. electric motor   B. transformer
   C. power grid   D. controller
2. Controllers can cause a wide range of dynamic systems behave _____.
   A. in a good manner   B. in a bad manner
   C. in a desired manner   D. in an unexpected manner
3. A _____ is not an electronic component.
   A. motor   B. diode
   C. transistor   D. capacitor
4. Discrete components have many disadvantages except _____.
   A. consuming much space   B. consuming much time
   C. consuming much power   D. speed limitation
5. The design of instruments in instrumentation engineering requires a good understanding of _____.

A. mathematics B. physics
C. chemistry D. electromagnetic theory

6. For senior engineers managing a team of technicians, _____ are important.
A. written skills B. communication skills
C. documentation skills D. project management skills

II. Fill in the blanks with proper words according to the passage.

1. Power engineering deals with the _____, _____, and _____ of electricity.
2. Power engineers work on the _____ and _____ of the power grid as well as the power systems that connect to it.
3. _____ control is widely used in many modern automobiles.
4. Since the mid-to-late 1950s, the term electronic engineering gradually took the place of the term _____.
5. Telecommunications engineering focuses on the transmission of information across a _____.
6. Popular analog modulation techniques include _____ and _____.

III. Give brief answers to the following questions.

1. What is electrical engineering?
2. What are the wide applications of control engineering? Give at least two examples.
3. What are the advantages of integrated circuits?
4. What does instrumentation engineering deal with?

Vocabulary Building

IV. Fill in the table below by giving the corresponding translation.

English	Chinese
amplitude modulation	
cruise control	
	电容器

续上表

English	Chinese
	组件,元件
integrated circuit	
power surge	
	家用电器,家电
	维护,保养
variable	

V. Match the items listed in the following twocolumns.

1. sensor   a. height above sea-level
2. counterpart   b. system of electric-power cables for distributing power evenly over a large area
3. altitude   c. machine for producing electrical energy
4. blackout   d. person or thing that corresponds to or has the same function as sb. or sth. else
5. grid   e. a device that receives a signal or stimulus (as heat, pressure, light, motion, etc.) and responds to it in a distinctive manner
6. transformer   f. apparatus for stepping up or down the voltage of an electric power supply
7. generator   g. darkness resulting from the extinction of lights

✎ Translation Practice

VI. Translate the following paragraph from the passage into Chinese.

Power engineering deals with the generation, transmission, and distribution of electricity as well as the design of a range of related devices. These include transformers, electric generators, electric motors, high voltage engineering, and power electronics. In many regions of the world, governments maintain an electrical network called a power grid that connects a variety of generators together with users of their energy. Users purchase electrical energy from the grid, avoiding the costly exercise of having to generate their own. Power engineers may work on the design and maintenance of the power grid as well as the power systems that connect to it. Such systems are called on-grid power systems and may supply the grid with additional power, draw power from the grid, or do both. Power engineers may also work on systems that do not connect to the grid, called off-grid power systems, which in some cases are preferable to on-grid systems. The future includes satellite controlled power systems, with feedback in real time to prevent power surges and prevent blackouts.

Section C
Introduction to Mechanical Engineering

Mechanical engineering is the branch of engineering that deals with machines and the production of power. It is particularly concerned with forces and motion.

History of Mechanical Engineering

The invention of the steam engine in the latter part of the 18th century, providing a key source of power for the Industrial Revolution, gave an enormous impetus to the development of machinery of all types. As a result, a new major classification of engineering, mechanical engineering separated from civil engineering and dealt with tools and machines, developed and received formal recognition in 1847 in the founding of the Institution of Mechanical Engineers in Birmingham, England.

Mechanical engineering has evolved from the practice by the mechanic of an art based largely on trial and error to the application by the professional engineer of the scientific method in research, design, and production.

The demand for increased efficiency, in the widest sense, is continually raising the quality of work expected from a mechanical engineer and requiring of him a higher degree of education and training. Not only must machines run more economically but capital costs also must be minimized.

Fields of Mechanical Engineering

Development of machines for the production of goods The high material standard of living in the developed countries owes much to the machinery made possible by mechanical engineering. The mechanical engineer continually invents machines to produce goods and develops machine tools of increasing accuracy and complexity to build the machines.

The principal lines of the development of machinery have been an increase in the speed of operation to obtain high rates of production, improvement in accuracy to obtain quality and economy in the product, and minimization of operating costs. These three requirements have led to the evolution of complex control systems.

The most successful production machinery is that the mechanical design of the machines is closely integrated with the control system, whether the latter is mechanical or electrical in nature. A modern transfer line (conveyor) for the manufacture of automobile engines is a good example of the mechanization of a complex series of manufacturing processes. Developments are in hand to automate production machinery further, using computers to store and process the vast amount of data required for manufacturing a variety of components with a small number of versatile machine tools. One aim is a completely automated machine shop for batch production, operating on a three-shift basis but attended by a staff for only one shift per day.

Development of machines for the production of power　　Production machinery presupposes an ample supply of power. The steam engine provided the first practical means of generating power from heat to augment the old sources of power from muscle, wind, and water. One of the first challenges to the new profession of mechanical engineering was to increase thermal efficiencies and power; this was done principally by the development of the steam turbine and associated with large steam boilers. The 20th century has witnessed a continued rapid growth in the power output of turbines for driving electric generators, together with a steady increase in thermal efficiency and reduction in capital cost per kilowatt of large power stations. Finally, mechanical engineers acquired the resource of nuclear energy, whose application has demanded an exceptional standard of reliability and safety involving the solution of entirely new problems. The control systems of large power plants and complete nuclear power stations have become highly sophisticated networks of electronic, fluidic, electric, hydraulic, and mechanical components, all of which involve the province of the mechanical engineer.

The mechanical engineer is also responsible for the much smaller internal combustion engines, both reciprocating (gasoline and diesel) and rotary (gas-turbine and Wankel) engines, with their widespread transport applications. In the transportation field generally, in air and space as well as on land and sea, the mechanical engineer has created the equipment and the power plant, collaborating increasingly with the electrical engineer, especially in the development of suitable control systems.

Development of military weapons　　The skills applied to war by the mechanical engineer are similar to those required in civilian applications, though the purpose is to enhance destructive power rather than to raise creative efficiency. The demands of war have channeled huge resources into technical fields, however, and led to developments that have profound benefits in peace. Jet aircraft and nuclear reactors are notable examples.

Bioengineering　　Bioengineering is a relatively new and distinct field of mechanical engineering that includes the provision of machines to replace or augment the functions of the human body and of equipment for use in medical treatment. Artificial limbs have been developed incorporating such lifelike functions as powered motion and touch feedback. Development is rapid in the direction of artificial spare-part surgery. Sophisticated heart-lung machines and similar equipment permit operations of increasing complexity and permit the vital functions in seriously injured or diseased patients to be maintained.

Environmental control Some of the earliest efforts of mechanical engineers were aimed at controlling man's environment by pumping water to drain or irrigate land and by ventilating mines. The ubiquitous refrigerating and air-conditioning plants of the modern age are based on a reversed heat engine, where the supply of power "pumps" heat from the cold region to the warmer exterior.

Many of the products of mechanical engineering, together with technological developments in other fields, have side effects on the environment and give rise to noise, the pollution of water and air, and the dereliction of land and scenery. The rate of production, both of goods and power, is rising so rapidly that regeneration by natural forces can no longer keep pace. A rapidly growing field for mechanical engineers and others is environmental control, comprising the development of machines and processes that will produce fewer pollutants and of new equipment and techniques that can reduce or remove the pollution already generated.

Functions of Mechanical Engineering

Four functions of the mechanical engineering, common to all the fields mentioned, are cited. The first is the understanding of and dealing with the bases of mechanical science. These include dynamics, concerning the relation between forces and motion, such as in vibration; automatic control; thermodynamics, dealing with the relations among the various forms of heat, energy, and power; fluid flow; heat transfer; lubrication; and properties of materials.

Second is the sequence of research, design, and development. This function attempts to bring about the changes necessary to meet present and future needs. Such work requires not only a clear understanding of mechanical science and an ability to analyze a complex system into its basic factor, but also the originality to synthesize and invent.

Third is the production of products and power, which embraces planning, operation, and maintenance. The goal is to produce the maximum value with the minimum investment and cost while maintaining or enhancing longer term viability and reputation of the enterprise or the institution.

Fourth is the coordinating function of the mechanical engineering, including management, consulting, and, in some cases, marketing.

In all of these functions there is a long continuing trend toward the use of scientific instead of traditional or intuitive methods, an aspect of the ever-growing professionalism of mechanical engineering. Operation research, value engineering, and PABLA (problem analysis by logical approach) are typical titles of such new rationalized approaches. Creativity, however, cannot be rationalized. The ability to take the important and unexpected step that opens up new solutions remains in mechanical engineering, as elsewhere, largely a personal and spontaneous characteristic.

The Future of Mechanical Engineering

The number of mechanical engineers continues to grow as rapidly as ever, while the duration and quality of their training increase. There is a growing awareness, however, among engineers and in the community at large that the exponential increase in population and living standards is raising formidable problems in pollution of the environment and the exhaustion of natural resources; this clearly heightens the need for all of the technical professions to consider the long-term social effects of discoveries and developments. There will be an increasing demand for mechanical engineering skills to provide for man's needs while reducing to a minimum the consumption of scarce raw materials and maintaining a satisfactory environment.

(1350 words)

Quoted from *English for Machinery*

by Bu Yukun

New Words and Expressions

augment [ɔːgˈment] v.	增加,增大
bioengineering [ˌbaiəuˌendʒiˈniəriŋ] n.	生物工程(学)
collaborate [kəˈlæbəreit] v.	合作,协作
dereliction [derəˈlikʃn] n.	玩忽职守;抛弃,遗弃
exponential [ˌekspəˈnenʃl] n. & a.	指数 & 指数的
formidable [ˈfɔːmidəbl] a.	强大的,令人敬畏的
heart-lung n.	人工心肺机
hydraulic [haiˈdrɔlik] a.	液压的,水力的
impetus [ˈimpitəs] n.	推动力,冲力;促进
irrigate [ˈirigeit] v.	灌溉
lubrication [ˌluːbriˈkeiʃn] n.	润滑,润滑作用
presuppose [ˌpriːsəˈpəuz] v.	以……为先决条件;假定,预料
province n.	(学问、活动或责任的)范围
reciprocating [riˈsiprəˌkeitiŋ] a.	往复的
reduction [riˈdʌkʃn] n.	减少
rotary [ˈrəutəri] a.	旋转的,绕轴转动的
spare-part n.	零部件

spontaneous [spɔnˈteiniəs] a.	自发的,自然的
synthesize [ˈsinθəsaiz] v.	合成,综合
thermodynamics [ˌθəːməudaiˈnæmiks] n.	热力学
ubiquitous [juːˈbikwitəs] a.	普遍存在的,无所不在的
ventilate [ˈventileit] v.	使通风
viability [ˌvaiəˈbiləti] n.	可行性
vibration [vaiˈbreiʃn] n.	振动
batch production	批量生产
capital cost	基建费,投资费
operation research	运筹学
value engineering	价值工程,工程经济学

Notes

1. PABLA (problem analysis by logical approach)　　　逻辑法问题分析

📖 Text Understanding

Ⅰ. **Choose the best answer according to the passage.**

1. Mechanical engineering is particularly concerned with ＿＿＿＿.
   A. stress and strains　　　　　　B. forces and motion
   C. power　　　　　　　　　　　D. production of machines

2. ＿＿＿＿ provided a key source of power for the Industrial Revolution.
   A. Turbine　　　　　　　　　　B. Motors
   C. Steam engine　　　　　　　　D. Electric power

3. The fields of mechanical engineering include the following except ＿＿＿＿.
   A. the development of machines　　B. the development of computers
   C. the development of weapons　　D. environmental control

4. In the following four choices, ＿＿＿＿ is NOT the requirements that have led to the evolution of complex control systems?
   A. the speed of operation　　　　B. high rates of production
   C. minimization of operation cost　D. higher degree of education

5. Mechanical engineers collaborate more and more with ＿＿＿＿ especially in the development of suitable control systems.

A. electrical engineers B. program designers
C. experienced technicians D. skilled workers

II. Fill in the blanks with proper words according to the passage.

1. Mechanical engineering deals with _____ and _____.
2. With the increased efficiency, machines should run more _____, and capital costs should be _____.
3. The mechanical design of a successful production machinery is closely integrated with _____.
4. Production machinery presupposes _____.
5. _____ and _____ are notable examples of the transformed application from military weapons to civilian use.

III. Give brief answers to the following questions.

1. What are the fields of mechanical engineering?
   _____
2. What does a completely automated machine shop look like?
   _____
3. What is bioengineering?
   _____
4. What are the functions of mechanical engineering?
   _____
5. What are the threats to mankind with the development of mechanical engineering?
   _____

Vocabulary Building

IV. Fill in the table below by giving the corresponding translation.

English	Chinese
	生物工程
exponential	
	润滑
	零部件
viability	
batch production	
	基建费,投资费
thermodynamics	

V. Fill in the blanks with the words from the passage. The first letter of the word is given.

1. The invention of the steam engine gave an enormous i_____ to the development of machinery of all types.
2. The mechanical engineer continually invents machines to produce goods and develops machine tools of increasing a_____ and complexity to build the machines.
3. He is an expert in mechanical engineering, and philosophy is not his p_____.
4. Mechanical engineers often c_____ with electrical engineers in developing suitable control systems.
5. The goal is to produce the m_____ value with the minimum investment and cost.

✎ Translation Practice

VI. Translate the following sentences from the passage into Chinese.

1. Mechanical engineering has evolved from the practice by the mechanic of an art based largely on trial and error to the application by the professional engineer of the scientific method in research, design, and production.

2. The control systems of large power plants and complete nuclear power stations have become highly sophisticated networks of electronic, fluidic, electric, hydraulic, and mechanical components, all of these involving the province of the mechanical engineer.

3. Creativity, however, cannot be rationalized. The ability to take the important and unexpected step that opens up new solutions remains in mechanical engineering, as elsewhere, largely a personal and spontaneous characteristic.

Translation Skill I

科技英语翻译学习的理论基础
——以系统功能语言学为例

我国"一带一路"倡议的实施与推进催生了大量对科技英语翻译的现实需求,科技英语翻译人才从数量到质量都呈现了显著增长,各类研究院所、工厂企业、出版单位、新闻媒体、政府机构等急需大批高质量的科技英语翻译人才。

合理的翻译理论对翻译实践具有解释功能,有助于译者翻译观念的形成。下文将结合系统功能语法探讨科技英语翻译的学习过程。Hatim & Mason(1990)指出,修辞功能是所有文本的特质,是文本生成者的综合意图,并通过文本功能来落实。科技英语的修辞功能为科技英语的交际行为提供规范和约束,并决定科技英语语篇的功能结构和翻译策略。

1. 翻译语境维度

Hatim & Mason(1990)从社会交际、符号学、语用学和认知等四维视角,解释了文本语言的现象和结构(图1):

1) 社会交际,即通过语域来分析翻译的社会交际功能。Hatim & Mason 依据系统功能模式和语域理论,提倡将文本的语言符号和非语言符号结合起来,放置在文本的情境语境中解读文本意义,以此阐述文本修辞功能与翻译的关系。

2) 符号学,即通过语言和非语言符号考察文本的衔接、连贯与互文功能。语言是符号模型系统,符号是人类交际互动的本质,是范畴化的连贯系统。

图1 翻译语境维度

3) 语用学,即通过言语行为保持翻译目的与意向。文本言语的顺序表征言语行为的内在相互关系,决定文本符号的推进模式和连贯性。

4) 认知,即通过认知模型、认知框架解读科技文本。

通过上述四维翻译语境透视和分析,译者根据科技英语语篇体裁,掌握其功能结构,通过英汉语言在词汇、句法及语篇层面的对比,探讨科技英语文本的翻译策略和方法。

2. 科技英语语篇翻译与功能结构

科技英语语篇翻译是特殊的语域(marked field of discourse),对于科技译者而言不仅是双语术语对等的问题,而且涉及科技文本作者和使用者的背景和目的,反映不同体裁的主题(语场,field)、科技交际者(语旨,tenor)和文本功能结构与语言形式(语式,mode)。高巍、范波(2020)根据语域将科技文本范畴化为表1。

科技英语语篇的范畴 表1

文本正式程度	语场:使用范围	语旨:交际者	语式:语言形式
高	科技论文、著作、报告	科学家之间	信息极为精确,术语较多,句法严密;概念翻译要精准。
中	产品说明书、操作手册	工程师、用户和商务人员之间	信息较为准确,句法简练,语言清晰;概念翻译要准确。
低	科普读物	科技专业人员与科普读者	信息较少,精确度较低,翻译简洁、生动、易懂。

科技英语翻译语篇的范畴是根据文本的正式程度由易到难、循序渐进,从低(科普读物)到中(产品说明书、操作手册),再到高(科技论文、著作和报告)。Seleskovitch & Ledere(1989)指出,科技英语翻译学习要从科普读物以及日常生活中含有一定科技原理的产品开始,包括电视、空调、汽车、飞机等。科技英语翻译不是系统学习科技知识,而是培养译者的好奇心,使其不断探索科技原理的关系。其次,Seleskovitch & Ledere(同上)还认为,科技英语的翻译内容广泛,但科技知识的原理是有限的,并具有普遍性。科技英语译者对翻译的科技主题应做到"懂",而不是"会"。

Trimble(1985)将科技修辞功能定义为科技英语语篇的组织结构和内容,修辞功能是语篇概念的一部分,进而将科技英语语篇功能划分为:1)描写(物理、功能、过程);2)定义(正式、半正式、非正式);3)分类(整体、局部、隐含、过程);4)指示。同时,他将科技文本的修辞方法划分为自然模式与逻辑模式,自然模式包括时间顺序、空间顺序、因果关系;逻辑模式包括主次顺序、比较和对比、类比、举例、图示与文本等关系。如表2所示。

科技英语语篇功能与结构 表2

语篇功能	
描写:表述对事物的认识和理解	定义:解释科技新概念、新技术和新方法
分类:按照种类、等级或性质归类	指示:系列操作过程
语篇结构	
自然模式:时间、空间、因果	
逻辑模式:主次顺序、比较和对比、类比、举例、图示与文本	

简言之,在科技英语翻译学习中,将翻译语境维度与科技英语语篇修辞功能、结构以及科技背景知识相结合有助于有效分析和翻译科技英语的语篇和文本。

摘选自高巍、范波(2020)
《科技英语翻译教学再思考:理论、途径和方法》

PART II
Electrical Engineering

Unit 2

Electricity

💬 Before-Class Reading

Answer the following question before reading the passage:

1. Do you know the famous experiment "kite in a thunderstorm" conducted by Benjamin Franklin? If you do, describe the experiment in your own words; if not, consult the relevant information including its significance in the study of electricity, and share it with your classmates.

Section A
Electricity in Early Days

Ancient Developments

Long before any knowledge of electricity existed, people were aware of shocks from electric fish. Ancient Egyptian texts dating from 2750 BCE referred to these fish as the "Thunderer of the Nile", and described them as the "protectors" of all other fish. Electric fish were again reported millennia later by ancient Greek, Roman and Arabic naturalists and physicians. Several ancient writers attested to the numbing effect of electric shocks delivered by electric catfish and electric rays, and knew that such shocks could travel along conducting objects. Patients suffering from ailments such as gout or headache were directed to touch electric fish in the hope that the powerful jolt might cure them. Possibly the earliest and nearest approach to the discovery of the identity of lightning, and electricity from any other source, is to be attributed to the Arabs, who before the 15th century had the Arabic word for lightning applied to the electric ray.

18th-Century Developments

In the 18th century, Benjamin Franklin conducted extensive research in electricity, selling his possessions to fund his work. In June 1752 he was reputed to have attached a metal key to the bottom of a dampened kite string and flown the kite in a storm-threatened sky. A succession of sparks jumping from the key to the back of his hand showed that lightning was indeed electrical in nature. He also explained the apparently paradoxical behavior of the Leyden jar as a device for storing large amounts of electrical charge, by coming up with the single fluid, two states theory of electricity.

As for the Leyden jar, it is a kind of condenser invented about the middle of the 18th century. An early type consisted merely of a bottle containing water (or alcohol or mercury) with a long nail dipping into the liquid. The use of the Leyden jar was that it enabled electricity to be stored. True, it was not of much practical value, but at least it caused plenty of amusement. We read accounts of people experimenting with electric shocks from Leyden jars. Indeed, on one occasion, a whole company of monks formed a line nine hundred feet long with wires making contact from man to man. They certainly experienced more than they expected; for when the ends of the wires were joined to a battery of charged Leyden jars, all the monks leaped into the air.

In 1791, an Italian, Luigi Galvani, discovered by chance that electricity from a Leyden jar or an electric machine would cause the legs of dead frog to move. This puzzled Galvani, so he started experimenting in order to try to discover the cause. He discovered that if two pieces of unlike metal, say copper and zinc, were placed on a nerve and a muscle, and then brought into contact with each other, the dead frog could be made to move. Galvani realized that it was the action of two different metals in the frog's leg that produced the electricity.

Galvani's discovery greatly interested one of his countrymen, Alessandro Volta, who carried out many experiments with metals and chemicals. Volta succeeded in producing a simple form of cell which is the parent of modern batteries. The invention of Volta's cell was important, because it meant that for the first time it was possible to maintain an electric current for a long period. This was unlike the Leyden jar, which merely gave a sudden rush of current.

19th-Century Developments

In 1819 the Danish scientist Oersted, like many other great pioneers, found out accidentally that a relationship exists between electricity and magnetism. By chance a wire connected to a battery was moved near a compass needle on the table. Oersted was astonished to notice that when current passed through the wire, the compass needle was moved. It no longer pointed to the north. Oersted's discovery interested the French scientist Ampere, and as a result of patient experimenting, he succeeded in fixing the law which governs electromagnetic effects.

We now come to the man whose discoveries made modern electrical engineering possible. He was Michael Faraday, born in 1791 in Yorkshire. Faraday conducted his memorable experiments in 1831, and discovered that an electric current will flow in a coil of wire when the coil is rotated between the poles of a magnet. Basic generators in power stations today work on this principle.

By and large, the discovery of electricity and the developments of electrical engineering are not realized in a single day or by a single scientist. The history of electrical engineering reflects the perseverance, the dedication, and the contribution of scientists worldwide.

(768 words)

https://encyclopedia.thefreedictionary.com/History+of+electrical+engineering

New Words and Expressions

ailment ['eilmənt] n.	病痛,小病
coil [kɔil] n.	线圈,绕组
compass ['kʌmpəs] n.	指南针
condenser [kən'densə] n.	冷凝器,电容器
conduct v.	导电,导热
gout [gaut] n.	痛风(病)
jolt [dʒəult] n. & v.	震动,颠簸,摇晃
magnet ['mægnət] n.	磁铁,磁石
magnetism ['mægnətizəm] n.	磁性,磁力;吸引力
mercury ['mə:kjəri] n.	水银,汞
numbing ['nʌmiŋ] a.	令人麻木的,使人失去知觉的
paradoxical [ˌpærə'dɔksikl] a.	似自相矛盾的
zinc [ziŋk] n.	锌
electric ray	电鳐
electrical charge	电荷
electromagnetic effect	电磁效应
Leyden ['laidn] jar	莱顿瓶

Exercises

📖 **Text Understanding**

Ⅰ. Decide whether the following statements are true (T) or false (F) according to the passage.

1. Some ancient writers reported that touching electric catfish would give people the

numbing feeling.
2. In a "kite in a thunderstorm" experiment, Franklin attached a metal key to the bottom of a dry kite string and flew the kite in a storm-threatened sky.
3. The Leyden jar was of much value in practice.
4. Both Galvani and Volta are Italian scientists.
5. Faraday discovered that an electric current will flow in a coil of wire when the coil is rotated between the poles of a magnet.

II. Fill in the blanks with proper words according to the passage.
1. Ancient Egyptian referred to electric fish as the "_____".
2. The "kite in a thunderstorm" experiment proved that lightning was _____ in nature.
3. Volta succeeded in _____ which is the parent of modern batteries.
4. Oersted found out by chance that a relationship exists between _____ and _____.
5. Ampere succeeded in fixing the law which governs _____.

III. Give brief answers to the following questions.
1. What did a Leyden jar look like?

2. Why was the invention of Volta's cell important?

3. What was Oersted's discovery?

Vocabulary Building

IV. Fill in the table below by giving the corresponding translation.

English	Chinese
	冷凝器
magnetism	
	电荷
	电磁效应
coil	

Translation Practice

V. Translate the following paragraph into English.

闪电是自然界常见的电现象之一。以前人们对闪电知之甚少,只是对其可怕的破坏感到恐惧。富兰克林的著名实验"雷电中的风筝"揭开了人们认识闪电的新篇章。随着电学研究的深入,人们知道了闪电与积雨云(cumulonimbus)中大量电量的突然释放有关。现在人们已能很好地保护自己免遭闪电的袭击。

In-Class Reading

Section B
Electrons and Electricity

More than two thousand years ago the Greek philosopher Thales observed that when a piece of amber, a hardened gum from trees, was rubbed with a material like wool or fur, it attracted certain other kinds of material. This ability to attract (and also to repel, as it was later discovered) other objects is due to electric charge. The phenomenon itself came to be called static electricity. "Electricity" comes from the Greek word for "amber"; "static" indicates that the charge remains stationary, that is, it remains bound to the material that has been charged.

It was many hundreds of years before any further significant observations were made about the phenomenon of static electricity. Then it was discovered that many other materials besides amber could be charged by rubbing, which produced friction. A more important discovery was that there were two kinds of electrical charges.

These two kinds of charges were called positive and negative. A positive charge was indicated by a plus sign (+) and a negative charge by a minus sign (-). These symbols are still in universal use today. It was also discovered that like charges—two positive charges or two negative charges— repelled each other, whereas unlike charges—a positive and a negative charge—attracted each other.

Much later it was learned that the movement of tiny particles of matter called electrons caused electricity (Fig. 2-1). The electron is one of the particles that make up atoms, the basic units of matter of which a chemical element is composed. The center of the atom is a nucleus which contains almost the entire weight or mass of the atom. The nucleus itself consists of two different kinds of particles, protons and neutrons.

Electrons, which have only a very small mass in comparison to protons and neutrons, orbit at a very rapid speed around the nucleus, somewhat in the same manner as the earth and the other planets orbit around the sun which is illustrated in Fig. 2-2. Each atom contains an equal number of electrons and protons but may have a different number of neutrons. Each chemical element has been given an atomic number that equals the number of electrons or protons that the atom contains. Hydrogen, the lightest element occurring in nature, has atomic number 1, because an atom of hydrogen contains one electron and one proton. Uranium, one of the heaviest elements, has atomic number 92, for the 92 electrons and 92 protons contained in its atom. Copper, which plays an important part in electricity, has an atom containing 29 electrons and 29 protons, and thus atomic number 29.

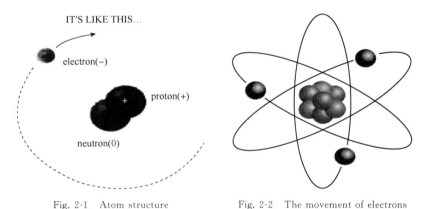

Fig. 2-1 Atom structure   Fig. 2-2 The movement of electrons

The electrons, as we have noted, is very light in weight and can be drawn out of its orbit around the much heavier nucleus. Electrons orbiting farther from the nucleus are those most easily drawn away. Orbiting electrons are called planetary electrons, and those that have been pulled away are known as free electrons. The electron has a negative electric charge, whereas the proton has a positive electric charge. The neutron has net charge. Drawing electrons away from the atom causes it to have a net positive electric charge because of the excess of protons. The atom then attracts the negatively charged free electrons.

Some materials permit the movement of free electrons more easily than others. These materials are called conductors—copper, silver, and aluminum are good examples. Other materials restrict the movement of free electrons. These are called insulators—and glass, rubber, and air are examples of them.

The movement of electrons requires energy, which can be defined as the ability to perform work. A considerable amount of energy is needed to produce the flow of electricity which can be seen in the sparks that jump between oppositely charged materials. In

Benjamin Franklin's famous experiment with a kite in a thunderstorm, he demonstrated that lightning was an electrical phenomenon involving an enormous amount of energy.

Two Italians, Luigi Galvani and Alessandro Volta, made it possible to produce electricity in a form which could be used. Galvani discovered that a frog's leg could be made to twitch when it touched two dissimilar pieces of metal. Shortly afterwards, Volta showed that the movement in the frog's leg was caused by a flow of current, meaning a flow of free electrons. The current was produced by the electrochemical properties of the two pieces of metal. As a result, he was able to invent the Voltaic cell, a chemical method of generating electricity. The term battery is often used to designate a cell, but more accurately, a battery is a group of cells connected together to produce a greater amount of electricity. An automobile has a battery which contains a number of cells as is shown in Fig. 2-3.

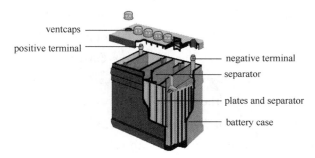

Fig. 2-3 Automobile battery

Voltaic cell (Fig. 2-4) consisted of strips of two different metals that were placed in a solution of salt water. One of the metals was more chemically active than the other and gave off electrons, which were attracted by the less active metal. The two pieces of metal could be externally connected by a wire to create a circuit—a path through which an electric current passes. The more active metal became the negative terminal of the circuit, and the less active metal became the positive terminal.

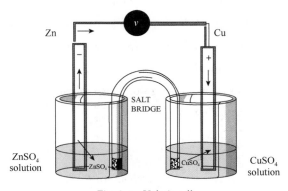

Fig. 2-4 Voltaic cell

Cells and batteries are still used as a source of electricity for many different purposes. A variety of metals and other substances are placed in solutions of salts, acids, or alkalis to make modern batteries. An automobile storage battery, for example, uses a solution of sulfuric acid in water with plates of lead and lead peroxide. A flashlight dry cell has a casing of zinc, the negative terminal, in which there is a paste of water and salts called ammonium chloride. The positive terminal is a carbon strip placed inside the paste.

The two plates of the cell are called electrodes, and are placed in a solution known as the electrolyte. The positive plate is the cathode, and the negative is the anode. A problem that arises in the use of cells is a chemical reaction that deposits hydrogen bubbles on the anode, preventing electrons from passing into the electrode. In some kinds of cells, the cathode will also gradually dissolve as a result of the continuous chemical reaction with the electrolyte. In storage batteries like those used in automobiles, the chemical reactions can be reversed by passing a reversed electric current through the electrolyte. This process is known as recharging, and it lengthens the life of the battery considerably.

The Voltaic cell opened up the way for the practical application of electricity. Chemical electricity is also widely used in the electrolytic process, or electrolysis. This process separates atoms that are combined in some molecules. A molecule is another tiny particle of matter which is made up of a combination of atoms. A molecule of water, for example, consists of two atoms of hydrogen and one of oxygen; a molecule of ordinary table salt is composed of one atom of sodium and one of chlorine.

When salt is dissolved, the chlorine atoms become negatively charged and the sodium atoms become positively charged because they have gained or lost electrons. Such electrically charged atomic particles are called ions. When electrodes are placed in the solution, the negatively charged chlorine ions collect at the anode, while the positively charged sodium ions collect at the cathode. There are other molecules which can be broken down into ionized atoms in solution. This is done by passing an electric current through the solution. This attracts ions of one element to the cathode and ions of the other to the anode.

The electrolytic process is used in refining aluminum from bauxite ore and in separating magnesium from chemicals found in sea water. It is also used for electroplating substances with metals such as silver and gold. Electrolysis is one of the most important processes available to chemists today.

(1340 words)

Quoted from *English for Electrical and Electronic Science and Technology*
by Bu Yukun

New Words and Expressions

acid ['æsid] n. 酸
alkali ['ælkəlai] n. 碱
aluminum [ə'lu:minəm] n. 铝
amber ['æmbə] n. 琥珀,琥珀色
anode ['ænəud] n. 阳极
bauxite ['bɔ:ksait] n. 铝土矿,铝矾土
cathode ['kæθəud] n. 阴极
chlorine [klɔ:ri:n] n. 氯
conductor n. 导体
electrode [i'lektrəud] n. 电极
electrolysis [i,lek'trɔləsis] n. 电解
electrolyte [i'lektrəulait] n. 电解液,电解质
electron [i'lektrɔn] n. 电子
electroplate v. 电镀
gum [gʌm] n. 黏胶,树胶
hydrogen ['haidrədʒən] n. 氢
insulator ['insəleitə] n. 绝缘体
ion ['aiən] n. 离子
ionize ['aiənaiz] v. (使)电离,(使)成离子
magnesium [mæg'ni:ziəm] n. 镁
molecule ['mɔlikju:l] n. 分子
neutron ['nju:trɔn] n. 中子
nucleus ['nju:kliəs] n. 原子核
ore [ɔ:] n. 矿石,矿砂
plate n. 金属板,厚钢板,板材
proton ['prəutɔn] n. 质子
separator n. 挡板,隔板
sodium ['səudiəm] n. 钠
solution n. 溶液
terminal ['tə:minl] n. 极,端子,终端
twitch [twitʃ] v. 抽动,颤动
uranium [ju'reiniəm] n. 铀(放射性化学元素)
ventcap n. 透气盖

ammonium chloride [əˈməuniəm ˈklɔːraid]	氯化铵
atomic number	原子序数
carbon strip	碳棒
electrochemical property	电化学性能
electrolytic [iˌlektrəˈlitik] process	电解过程
lead peroxide [pəˈrɔksaid]	过氧化铅
net charge	净电荷
planetary electron	轨道电子
sulfuric [sʌlˈfjuərik] acid	硫酸
Voltaic [vəˈteiik] cell	伏打电池

📖 Text Understanding

I. Choose the best answer according to the passage.

1. Generally speaking, an atom consists of ＿＿＿＿.
    A. electrons, protons, and nucleus
    B. electrons, protons, and neutrons
    C. electrons, free electrons, and planetary electrons
    D. positive charges and negative charges

2. In an atom, ＿＿＿＿ have no electric charge.
    A. electrons   B. protons
    C. planetary electrons   D. neutrons

3. The atomic number of a chemical element is counted by the number of ＿＿＿＿ that the atom contains.
    A. electrons or protons   B. electrons plus protons
    C. nucleus   D. neutrons

4. Some materials like ＿＿＿＿ restrict the movement of free electrons.
    A. copper   B. silver
    C. steel   D. rubber

5. Electricity can perform work because the movement of free electrons transmits ＿＿＿＿.
    A. excess of protons   B. electric charges
    C. energy   D. mass

6. When free electrons are flowing through a conductor, an electric ＿＿＿＿ has been

established.

A. route B. circle
C. river D. circuit

7. The _____ is an electrode with a negative charge.

A. anode B. cathode
C. peroxide D. bubble

8. An atom that has lost or gained electrons so that it is electrically charged is called a(n)_____.

A. electrolysis B. sodium
C. chlorine D. ion

II. Fill in the blanks with proper words according to the passage.

1. The ability to attract or repel other objects is due to _____.
2. Two positive or two negative charges _____ each other, while a positive and a negative charge _____ each other.
3. The nucleus consists of two different kinds of particles, namely _____ and _____.
4. Orbiting electrons are called _____, and those that have been pulled away are called _____.
5. Materials which restrict the movement of free electrons are called _____.
6. When salt is dissolved, the chlorine atoms become _____ and the sodium atoms become _____, because they have gained or lost electrons.

III. Give brief answers to the following questions.

1. How to define conductors? What are good examples of conductors?

2. What is the difference between a cell and a battery?

3. What are the two problems that arise in the use of cells or batteries?

4. What can be done to lengthen the life of some kinds of cells or batteries?

IV. Discussion.

Describe the different parts of an atom and their relationship to each other.

Vocabulary Building

V. Fill in the table below by giving the corresponding translation.

English	Chinese
	铝
	导体
electrolyte	
molecule	
	电化学性能
	轨道电子
nucleus	
separator	
	原子序数

VI. Match the items listed in the following two columns.

1. proton
2. electrode
3. neutron
4. electrolysis
5. insulator
6. electroplate

a. either of two solid conductors by which an electric current enters or leaves a battery
b. separation of a substance into its chemical parts by an electric current
c. cover sth. with a thin layer of metal
d. particle carrying no electric charge
e. particle with a positive electric charge
f. a material such as glass or porcelain with negligible electrical or thermal conductivity

Translation Practice

VII. Translate the following sentences from the passage into Chinese.

1. The movement of electrons requires energy, which can be defined as the ability to perform work. A considerable amount of energy is needed to produce the flow of electricity which can be seen in the sparks that jump between oppositely charged

Electrical Engineering PART 2

materials.

2. Voltaic cell consisted of strips of two different metals that were placed in a solution of salt water. One of the metals was more chemically active than the other and gave off electrons, which were attracted by the less active metal. The two pieces of metal could be externally connected by a wire to create a circuit—a path through which an electric current passes.

3. In storage batteries like those used in automobiles, the chemical reactions can be reversed by passing a reversed electric current through the electrolyte. This process is known as recharging, and it lengthens the life of the battery considerably.

Section C
Electricity and Electronics

Volta made his experiment cell in 1800, producing for the first time a steady, reliable electric current. During the 19th century, the development of practical applications of electrical energy advanced rapidly. The first major uses of electricity were in the field of communications—first for the telegraph and then the telephone. They used not only electric current but also electromagnetic effects.

Thomas Edison's invention of the electric light bulb, based on incandescent light, was perhaps the most momentous development of all, but not because it was such a unique invention. Actually, other people were working simultaneously on the same technical problem, and Edison's claim to the invention was disputed. It was momentous because it led to the creation of an electric power system which has since reached into nearly every corner of the world. Perhaps Edison's most important claim to fame is his pioneering work in engineering, which helped to provide electric service for New York City in 1882.

The applications of electricity have grown to the point where most of us lead "electrified lives", surrounded by a variety of devices that use electric energy. Less visible, but probably more important, are the thousands of ways industry has put electric energy to work.

It is quite remarkable that so much of this rapid development of electrical devices and the resulting industry took place during the 19th century, when the nature of electricity was not completely understood. We have already observed that for a long time, it was incorrectly believed that current flowed from positive to negative. It was not until 1897 that the British scientist Sir Joseph Thomson published a paper announcing his discovery of a subatomic particle, the electron (first called a corpuscle). Up to that time it had been generally believed that the atom was an indivisible particle of matter.

Thomson's discovery led to further experimentation into the structure of the atom. He may be considered the founder of the modern science of nuclear physics. Within the field of electricity, his work led to the creation of the science of electronics. There is so much confusion in current usage between the terms "electricity" and "electronics" that we should attempt to make some sort of distinction between them.

Electricity (Fig. 2-5) generally refers to the flow of free electrons through a conductor, in other words, to a current of electricity. The term includes the electric power supplied by generators and the distribution systems which transmit it to homes, offices, and factories.

Electronics (Fig. 2-6), on the other hand, deals with the movement of free electrons in a vacuum or in semiconductors. When the term first came into use, it referred to the behavior of free electrons in vacuum tubes like those used to transmit or detect radio waves. Since then it has been extended to include the movement of electrons in gases, liquids, and solids which had not previously been considered to be conductors.

Fig. 2-5 Electricity

Fig. 2-6 Electronics

Electronic refinements have greatly extended the uses and capabilities of some of the older electrical devices. The switching devices necessary for the direct dialing of telephone calls are the result of electronic engineering. The transistor, an invention which has revolutionized the science of electronics, was first developed for use in telephone equipment.

Regardless of the distinction made between the two fields, both must be understood by today's electrical and electronic engineers. Even an engineer working on the design of the newest computer must have a knowledge of circuits and electromagnetic effects. Electricity and electronics are really indivisible; each one forms part of the other.

There has been so much emphasis on electronic developments—and so much publicity given them—that until recently they seemed to have taken over the electrical and electronic field completely. The general attitude was that electric power systems had already advanced as far as they could, and that anything new and exciting would come from the area of electronics. However, the current shortage of energy sources throughout the world has shown that there is still a need for research and development in the field of electricity. This need exists primarily in three areas: improving present systems of generating and transmitting electric power, discovering practicable new systems for generating electric power, and creating systems to derive electricity from new sources.

We have noted that most electricity is generated by powerful electromagnetic devices turned by turbines that use water of steam as a source of power (Fig. 2-7). In recent years, however, nuclear energy derived from atomic fission, the splitting of the nuclei of atoms, has become more important. Research also continues into nuclear fusion, the joining together of atomic nuclei with a great release of energy. Finding a way to exploit this source of power would be a tremendous advance, since the fuel for fusion would consist principally of molecules that occur in water. Research of this kind is typically done by teams of scientists and engineers from a variety of different disciplines.

Fig. 2-7 Steam turbines

In order to reduce energy loss in the transmission of electricity, researchers are looking for methods of transmitting increasingly high voltages (Fig. 2-8). Transmission lines that will carry a million volts are presently under development. Systems for increasing the efficiency of transmitting direct current electricity have already been introduced. A direct current submarine line carrying 100 kilovolts (100,000 volts: a kilovolt equals a thousand volts) was put into operation in Sweden in 1954. The transmission of high-voltage direct current depends on an electronic device called a rectifier, which changes alternating current to direct current.

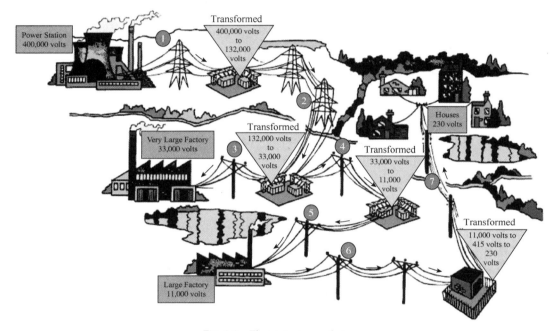

Fig. 2-8 Electricity transmission

Researchers are also trying to develop new systems of generating electricity, some of which involve new sources of energy. One part of this research has concentrated on finding a new source of power to drive the turbines, such as the kinetic energy (energy that comes from motion) of the wind and tides. Another line of research attempts to develop other known but so far impracticable methods for generating electricity. These include piezoelectricity—electricity that comes from pressure or weight applied to certain kinds of crystals. More promising is thermoelectricity, or the generation of electricity through heat. When the joint between two different metals is heated, a weak electromotive force is created. The joint is called a thermocouple, and several thermocouples joined in a series, like cells in a battery, increase the electromotive force. Thermoelectric generators, with heat supplied from radioactive materials, have been used in equipment for the space program. The action of light on some substances can also produce a release of electrons with an electromotive force. This effect is known as photoelectricity. It is familiar to most of us in the photoelectric cells that open and close automatic doors when a beam of light is broken.

Finally, attempts are being made to improve the means of producing electricity from chemical reactions in cells and batteries. Although electric cars powered by batteries (Fig. 2-9) have existed for a long time, they still cannot compete with cars powered by internal combustion engines, which burn increasingly scarce and expensive gasoline. There have been some promising discoveries in battery research, however, which may hasten the development of a practical battery-powered car. An electric car produced at a low enough price and with a long enough operation time would offer two advantages over the internal combustion engine: it would save fuel and avoid further pollution of the atmosphere.

Fig. 2-9　A charging electric car

(1241 words)

Quoted from *English for Electrical and Electronic Science and Technology* by Bu Yukun

New Words and Expressions

corpuscle ['kɔːpʌsl] n.	微粒；细胞
kilovolt ['kiləuvəult] n.	千伏
momentous [məu'mentəs] a.	重大的，重要的
nuclei ['njuːkliai] n.	原子核（nucleus 的复数形式）
photoelectricity n.	光电
piezoelectricity [paiiːzəuilek'trisiti] n.	压电
radioactive [ˌreidiəu'æktiv] a.	放射性的
rectifier ['rektifaiə] n.	整流器
refinement n.	改进，改善
semiconductor n.	半导体
submarine ['sʌbməriːn] a.	水下的，海底的
thermoelectricity ['θəːməuilek'trisəti] n.	热电

transmit [trænz'mit] v.	传输,传送,传递
turbine ['tə:bain] n.	涡轮机,汽轮机
vacuum ['vækju:m] n.	真空
alternating ['ɔ:ltəneitiŋ] current	交流电
atomic fission ['fiʃn]	原子裂变
direct current	直流电
electric power system	电力系统
electromotive force	电动势
incandescent [ˌinkən'desnt] light	白炽灯
internal combustion engine	内燃机
kinetic [ki'netik] energy	动能
nuclear fusion ['fju:ʒn]	核子融合
photoelectric cell	光电池,光电管
steam turbine	蒸汽涡轮

📖 Text Understanding

Ⅰ. Decide whether the following statements are true (T) or false (F) according to the passage.

1. The most momentous development of all was the invention of incandescent light.
2. Edison's claim to the invention of the electric light bulb was disputed.
3. In the 19th century, the nature of electricity was understood completely by scientists.
4. Current flowed from positive to negative was a misunderstanding.
5. The transistor was first developed for use in telegraph equipment.
6. Kinematic energy comes from motion.

Ⅱ. Fill in the blanks with proper words according to the passage.

1. Telegraph and telephone use not only _____, but also _____.
2. Sir Joseph Thomson's discovery led to further experimentation into _____.
3. _____ have greatly extended the uses and capabilities of some of the older electrical devices.
4. A _____ changes alternating current to direct current.
5. _____, with heat supplied from radioactive materials, have been used in

equipment for the space program.

6. The impracticable methods mentioned in the passage for generating electricity are _____, _____, and _____.

III. Discussion.

Try to give a brief summary of the forms of electricity.

_____
_____
_____
_____
_____

Vocabulary Building

IV. Fill in the table below by giving the corresponding translation.

English	Chinese
	光电
	压电
semiconductor	
turbine	
	交流电
	原子裂变
rectifier	
electromotive force	
	动能

V. Fill in the blanks with the words from the passage. The first letter of the word is given.

1. Edison's p_____ work in electricity research has made great contribution to the society.

2. It is r_____ that people have invented so many electrical devices in such a short time.

3. Conductors r_____ to those materials that permit the movement of free electrons more easily.

43

4. Electricity includes the electric power supplied by generators and the distribution systems which t_____ it to homes, offices, and factories.
5. It is quite prospective to e_____ the wind energy in Tibet.

✍ **Translation Practice**

Ⅵ. Translate the following paragraph into English.

　　第一辆展出的电动汽车在19世纪30年代被制造出来,而商用电车直到19世纪末才开始使用。然而,电动汽车没有像内燃机汽车那样获得巨大成功,因为内燃机汽车通常具有行驶里程长、重新加油容易等优势。今天对环境的关心,特别是关于噪声和废气排放,电池与燃料电池的新发展,逐步使得电动汽车的优势获得新的平衡。

💬 **Translation Skill II**

科技英语翻译的原则和标准

　　科技英语(English for Science and Technology,简称EST)作为一种重要的英语文体,与非科技英语文体相比,具有长句多、词义难、惯用被动语态、词性转换频繁、非谓语动词多、专业性强等特点。因此,科技英语的翻译也有别于其他英语文体的翻译,必须遵循特定的翻译原则和标准。具体如下:

1. 科技英语翻译原则

　　第一,在科技英语翻译实践中,译者必须了解和熟悉所要翻译的科技内容,如技术原理、科学知识、工程技术规范等。

　　第二,译者必须掌握相关科技内容涉及的专业术语。

　　第三,译者必须十分明确英语和汉语的共通之处,在汉译时对其共性可采取"直译",使得译文既忠实原文内容,又符合原文结构形式。

　　第四,译者还需注意英语和汉语的差异。翻译中既要避免因为直译而出现"英化汉语",即应该在忠实原文内容的前提下摆脱原文结构的束缚,使译文符合汉语表达规范;又要基于科技文献准确、严谨的特点,遵循"能直译处尽量直译,不能直译处采用意译"的原则。

2. 科技英语翻译标准

　　第一,忠实准确。科技英语以传递科技信息为目的,集中体现语言的信息功能,基本或完全不涉及个人情感和复杂的社会民族文化。正如Widdowson(1978)所说:"科技语篇代表一种将现实概念化的方式,一种为保持其科技属性而必须独立于不同语言、不同文化之外的交流方式。"因此,科技英语的译文必须忠实于原文,必须客观完整地表达、传递、重视原文的内容。

第二，通顺流畅。通顺流畅指的是译文文本流畅自如，译文在读者头脑中产生的印象和原文在读者头脑中产生的印象基本相同，译文读者能够产生跟原文读者基本相同的感受。如果违背这一要求，就会出现"翻译腔"。

第三，规范专业。翻译的规范专业要求译文的专业术语表述符合科技语言和术语的规范，尽可能利用已有的约定俗成的定义、术语和概念。科技文体是应用于科学技术领域进行沟通和交流的正式文体，其正式程度根据情景、语境等的不同而有所不同，正式程度越高，专业术语、定义和概念也就越多，文体的专业化程度也就越高，翻译时对译文的规范化、专业化要求也就越高。

总之，在翻译实践中译者必须全面理解并遵循以上翻译原则和标准，力求高质量地完成科技文献的翻译工作。

Unit 3

Power Source

Before-Class Reading

Answer the following questions before reading the passage:
1. What are the differences between direct current and alternating current?
2. Which kind of current is most commonly adopted in our ordinary life, direct current or alternating current?

Section A
Direct Current and Alternating Current

There are two kinds of current flows, direct current, usually abbreviated to d.c., and alternating current, a.c. (Fig. 3-1). In direct current, the flow of electrons moves steadily in one direction. This is the kind of current that is generated by a flashlight cell or an automobile battery. In alternating current, the flow of electrons is reversed rapidly over and over again. This is called electromagnetism.

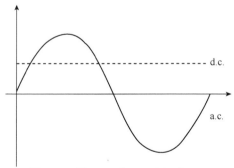

Fig. 3-1 The waveforms of d.c & a.c.

Originally, electrical energy that was used for commercial purposes was generated as direct current. But alternating currents supply systems rapidly replaced direct current ones, and today electrical energy is practically always generated and transmitted as alternating current. The change-over is fundamental due to the ease with which alternating current can be transformed to higher or lower voltages, thereby facilitating the transmission of high power over considerable lengths of line and reducing costs. Moreover, alternating current can be generated more cheaply than direct current with large units, and a.c. motors are usually preferred to d.c. motors for constant speed work. When direct current is necessary, as for examples in traction, electrolytic processes, variable speed units, etc., it is usual to convert alternating current to direct current by means of rectifiers.

To have a better understanding of alternating currents and voltages, it is desirable to begin with a consideration of the more general situation. A function of time is called "alternating" when, after a time interval of period, it repeats a previous succession of positive and negative values. In other words, such a function repeats itself exactly over equal intervals of time. So an alternating current is a rate of flow of electricity which does not have a constant value in time but grows to maximum value, decreases, changes its direction, reaches a maximum value in the new direction, returns to its original value and then repeats this cycle an indefinite number of times. The graphical representation of the variations of an alternating current plotted as a function of time is called the waveform of that current (Fig. 3-2). Generally the period is represented by T and measured in seconds. The reciprocal of this value T is called frequency and is defined as the number of periods occurring in the unit of time. We often express frequency in hertz or cycles per second, a cycle being complete series of changes taking place in one period of a periodically varying quantity.

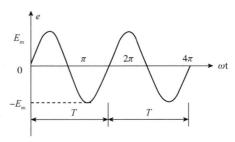

Fig. 3-2 The waveform of alternating current

The choice of the frequency value of a power system is both a technical and economic consideration. With the building of large transmission system, it has been necessary to standardize frequencies. In Europe, as well as in China, a frequency of 50 hertz has been widely adopted, whereas in most of America 60 hertz is used.

(470 words)

From *Fundamentals of Electrical Engineering*

by Johnson D.

New Words and Expressions

constant *n*.　　　　　　　　　　　　常数；恒量
frequency *n*.　　　　　　　　　　　　频率
hertz [həːts] *n*.　　　　　　　　　　　赫兹（声波频率单位）
reciprocal [riˈsiprəkl] *n*.　　　　　　　倒数
traction [ˈtrækʃn] *n*.　　　　　　　　　牵引，拖拉
a function of time　　　　　　　　　　时间函数

📖 Text Understanding

Ⅰ. **Decide whether the following statements are true (T) or false (F) according to the passage.**

1. A flashlight cell or an automobile battery is supplied with alternating current.
2. Commercial use of electrical energy is supplied with direct current.
3. A rectifier is capable of converting direct current to alternating current.
4. A function of time is called "alternating" if it repeats itself exactly over any intervals of time.
5. The reciprocal of value T is called frequency.

Ⅱ. **Fill in the blanks with proper words according to the passage.**

1. Today electrical energy is practically always generated and transmitted as _____.
2. In practical use, a.c. motors are usually preferred to d.c. motors for _____.
3. An alternating current is a rate of _____ which does not have a _____.
4. The graphical representation of the variations of an alternating current plotted as a function of time is called _____.
5. With the building of large transmission system, it has been necessary to _____.

Ⅲ. **Give brief answers to the following questions.**

1. What is d.c. and what is a.c.?

2. Why is electrical energy practically always generated and transmitted as alternating current?

3. What is frequency?

4. What is the standardized frequency in China's power supply?

Vocabulary Building

IV. Fill in the table below by giving the corresponding translation.

English	Chinese
	交流电
	直流电
a function of time	
reciprocal	
	频率
	常数

In-Class Reading

Section B
Electrical Power System

As the power industry grows, so do the economic and engineering problems connected with the generation, transmission, and distribution systems used to produce and handle the vast quantities of electrical energy consumed today. These systems together form an electrical power system (Fig. 3-3).

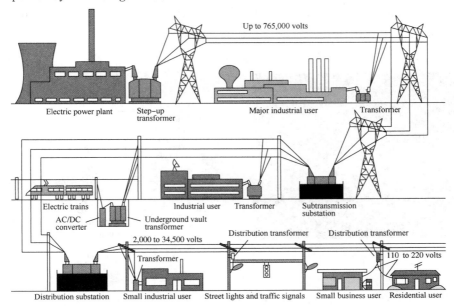

Fig. 3-3 Electrical power system

It is important to note that the industry that produces electrical energy is unique in that it manufactures its product at the very instant that it is required by the customer. Energy for the generation of electricity can be stored in the form of coal and oil, and of water in reservoirs and lakes, to meet future requirements, but this does not decrease the need for generator capacity to meet the customers' demands.

It is obvious that the problem of the continuity of service is very important for an electrical power system. No service can be completely protected from the possibility of failure and clearly the cost of the system will depend on its reliability. A balance must therefore be struck between reliability and cost; and the final choice will depend on the size of the load, its character, the source of possible interruptions, and the user's requirements. However, a net reliability gain is obtained by employing a certain number of generating units and by using automatic breakers for the separation into sections of the bus bars in generating stations and of the transmission lines in a national or international grid system. In fact, a large system comprises numerous generating stations and loads interconnected by high-capacity transmission lines. An individual unit of generation or set of transmission lines can usually cease to function without interrupting the general service.

The most usual system today for generation and for the general transmission of power is the three-phase system (Fig. 3-4 & Fig. 3-5). In favor of this system are its simplicity and its saving with respect to other a.c. system. In particular, for a given voltage between conductors, with a given power transmitted, with a given distance, and with a given line loss, the three-phase system requires only 75 per cent of the copper or aluminum needed in the single-phase system.

Fig. 3-4 Three-phase system

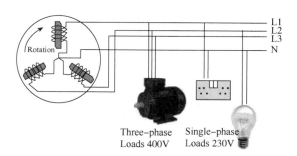

Fig. 3-5 Three-phase power

Another important advantage of the three-phase system is that three-phase motors are more efficient than single-phase ones. The sources of energy for large-scale electricity generation are:

1. steam obtained by means of a conventional fuel (coal, oil, or natural gas), the combustion of city refuse or the employment of nuclear fuel;
2. water;
3. diesel power from oil.

There are other possible sources of energy such as direct solar heat (Fig. 3-6), wind power (Fig. 3-7), tidal power, etc., but none of these has yet gone beyond the pilot-plant stage.

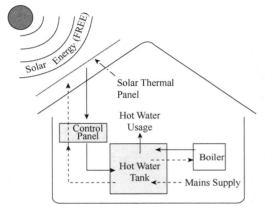

Fig. 3-6 Solar heat

Fig. 3-7 Wind power

In large steam power plants, the thermal energy stored in steam is converted into work by means of turbines. A turbine consists essentially of a shaft and a rotor fixed in bearings and enclosed in a cylindrical casing. The rotor is made to turn smoothly by means of jets of steam from nozzles around the periphery of the turbine cylinder. These steam jets strike blades attached to the shaft. Central power stations employ condensing turbines in which the steam passes into a condenser after leaving the turbine. Condensation is affected by the circulation of large quantities of cold water through the tubes of the condenser, thus increasing the expansion ratio of the steam and the consequent efficiency and work output of the turbine. The turbines are connected directly to large electricity generators.

In turbines the action of the steam is kinetic. There is progressive expansion of the steam from the high pressure and relatively small volume at which it enters the turbine to the low pressure and relatively very great volume at which it leaves.

Steam is made by heating water in a boiler. The usual boiler has a furnace in which fuel is burned, and the heat given off during combustion is conducted through the metal walls of the boiler to generate steam at a pressure within the boiler vessel. In nuclear plants, steam is generated with the aid of a reactor in which the controlled fission of uranium or plutonium supplies the necessary heat for the vaporization of water. Thus the reactor replaces the steam generator of conventional plants.

Use is made of the energy possessed by water in hydroelectric stations. In order to transform this energy into work, hydraulic turbines are used. Modern hydraulic turbines may be divided into two classes: impulse turbines and pressure or reaction turbines. Of the former, the Pelton wheel is the only type used in important installations; of the latter, the Francis turbine or one of its modifications is universally employed.

In an impulse turbine, the whole head of water is converted into kinetic energy before the wheel is reached, as the water is supplied to the wheel through a nozzle. In the pressure or reaction turbine the wheel (or runner) is provided with vanes into which water is directed by means of a series of guide vanes around the whole periphery. The water leaving these guide vanes is under pressure and supplies energy partly in the kinetic form and partly in the pressure form.

The diesel engine is an excellent prime mover for electricity generation in plant below about 10,000 KVA. It has the advantages of low fuel cost, a brief warming-up period and low standing losses. Moreover it requires little cooling water. Diesel generation is generally chosen for small power requirements by municipalities, hotels, and factories; hospitals often keep an independent diesel generator for emergency supply.

The transmission of electrical energy by means of lines (Fig. 3-8) is a great problem in electrical power systems. Transmission lines are essential for three purposes:

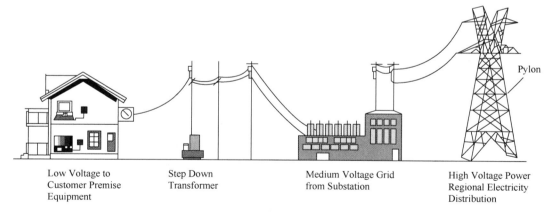

Fig. 3-8 The transmission of electrical energy

1. To transmit power from a hydroelectric site to a load center perhaps a considerable distance away;

2. For the bulk supply of power from steam stations to load centers a relatively short distance away;

3. For interconnection purposes to transfer energy from one system to another in case

of emergency.

The transmission voltage is determined largely by economic factors. In fact, in a transmission line, if the distance, the power, and the power loss are fixed, the total weight of the conductor varies inversely as the square of the transmission voltage. For the economic transmission of power over considerable distances, the voltage must therefore be high. Naturally with higher voltages, the insulation cost also rises; and to find the optimum voltage, we must strike a balance between this cost and the saving through the reduction of the cross-section of the conductors.

For high voltages, overhead-line construction is generally used with suspension-type insulators. Steel towers, called pylons (also see Fig. 3-8), serve to carry the insulators, with each conductor suspended from the bottom of a group or string of insulator units. The following types of conductors are those most commonly used: stranded copper conductors, hollow copper conductors, and ACSR (aluminum cable, steel reinforced) conductors.

Distribution includes all the parts of the electricity system between the power stations supplied from high-voltage transmission lines and the consumer's switch. Electric power is received from substations and distributed to the consumers at the voltage levels and with the degree of continuity that are acceptable to the various types of consumers. In large metropolitan systems, both overhead and underground distribution methods are used. Although underground distribution is more expensive than an overhead system, it is virtually necessary in heavily urbanized areas. In smaller towns and in the less congested districts of large cities, the entire distribution system is usually overhead.

(1281 words)

Quoted from *English for Electrical and Electronic Science and Technology*
by Bu Yukun

New Words and Expressions

bearing *n.*	轴承
blade [bleid] *n.*	(机器上旋转的)叶片,桨叶
boiler ['bɔilə] *n.*	锅炉
condensation [ˌkɔnden'seiʃn] *n.*	冷凝,凝结
cross-section *n.*	横截面
cylinder ['silində] *n.*	圆柱体;(发动机的)气缸
cylindrical [sə'lindrikl] *a.*	圆柱形的
diesel ['diːzl] *n.*	柴油

furnace [ˈfəːnis] n.	熔炉,火炉
gain n.	(电)增益
inversely [ˌinˈvəːsli] adv.	成反比地,相反地
load n.	负载,负荷
metropolitan [ˌmetrəˈpɔlitən] a.	大城市的,大都会的
municipality [mjuːˌnisiˈpæləti] n.	自治市
nozzle [ˈnɔzl] n.	喷嘴
optimum [ˈɔptiməm] a.	最佳的,最适宜的
periphery [pəˈrifəri] n.	边缘,周围
plutonium [pluːˈtəuniəm] n.	钚(放射性化学元素)
pylon [ˈpailən] n.	电缆塔,高压线铁塔
reservoir [ˈrezəvwɑː] n.	水库,蓄水池
rotor [ˈrəutə] n.	转子,转动部件
shaft [ʃɑːft] n.	主轴
substation [ˈsʌbsteiʃn] n.	变电站
vane [vein] n.	叶片,轮叶
vaporization [ˌveipəraiˈzeiʃn] n.	蒸发;汽化
automatic breaker	自动断路器
bulk supply	大批量供应
bus bar	汇流排,母线
city refuse [ˈrefjuːs]	城市垃圾
head of water	水头
hydroelectric station	水力发电站,水电站
impulse turbine	冲击式水轮机
pilot-plant stage	中试阶段
pressure or reaction turbine	压力或反应式水轮机
prime mover	原动力
power loss	功率损耗
single-phase system	单相系统
standing loss	标准损耗
steam jet	蒸汽喷射流
three-phase system	三相系统

Notes

1. ACSR (aluminum cable, steel reinforced) conductors　　ACSR(用钢加强的铝线)导线
2. stranded copper conductor　　多芯铜导线

3. the Francis turbine 弗朗西斯水轮机
4. the Pelton wheel 佩尔顿水轮机

Text Understanding

I. Choose the best answer according to the passage.

1. Electrical power system involves the following processes of power except _____.
   A. generation   B. transmission
   C. distribution   D. purchasing

2. The three-phase system is preferred to the single-phase system for the following reasons except _____.
   A. efficiency   B. usefulness
   C. simplicity   D. saving

3. The three-phase system saves _____ of the copper or aluminum than the single-phase system does.
   A. a quarter   B. a half
   C. three quarters   D. none

4. In nuclear plants, the _____ replaces the steam generator of conventional plants.
   A. turbine   B. reactor
   C. uranium   D. plutonium

5. The _____ is the only type of hydraulic turbines used in important installation.
   A. impulse turbine   B. pressure or reaction turbine
   C. Pelton wheel   D. Francis turbine

6. Below 10,000 KVA, _____ is an excellent prime mover for electricity generation in plant.
   A. the turbine   B. the generator
   C. the diesel engine   D. the internal combustion engine

II. Fill in the blanks with proper words according to the passage.

1. Today the most usual system for generation and transmission of power is _____.
2. The most typical examples of conventional fuels are _____, _____, and _____.

3. In large steam power plants, the thermal energy stored in steam is converted into work by means of _____.

4. In a power plant, the turbines are connected directly to _____.

5. Modern hydraulic turbines may be divided into two classes: _____ and _____.

6. For high voltages, overhead-line construction is generally used with _____.

III. Give brief answers to the following questions.

1. Why is the power industry unique?

2. What are the sources for large-scale electricity generation?

3. Why do central power stations use condensing turbines?

4. What is the difference of steam generating between conventional plants and nuclear ones?

5. What are the advantages of the diesel engine?

6. Where are overhead distribution and underground distribution respectively adopted? Why?

IV. Discussion.

Describe how the steam turbine propels the generation of electricity.

Vocabulary Building

V. Fill in the table below by giving the corresponding translation.

English	Chinese
	气缸
	变电站
city refuse	
cross-section	
	自动断路器
	大批量供应
pilot-plant stage	
hydroelectric station	
	三相系统
	柴油发动机

VI. Match the items listed in the following two columns.

1. shaft   a. machine or motor driven by a wheel which is turned by a current of water, steam, air or gas
2. reactor   b. bar or rod joining parts of a machine or transmitting power in a machine
3. bus bar   c. device reducing friction in part of a machine where another part turns
4. cross-section d. an electrical conductor that makes a common connection between several circuits
5. pylon   e. tall steel framework used for carrying overhead high-voltage electric cables
6. bearing   f. apparatus for the controlled production of nuclear energy
7. turbine   g. surface formed by cutting through sth., especially at right angles

✍ Translation Practice

VII. Translate the following sentences from the passage into Chinese.

1. It is obvious that the problem of the continuity of service is very important for an electrical power system. No service can be completely protected from the possibility of failure and clearly the cost of the system will depend on its reliability.

_____

_____

2. However, a net reliability gain is obtained by employing a certain number of generating units and by using automatic breakers for the separation into sections of the bus bars in generating stations and of the transmission lines in a national or international grid system.

3. In turbines the action of the steam is kinetic. There is progressive expansion of the steam from the high pressure and relatively small volume at which it enters the turbine to the low pressure and relatively very great volume at which it leaves.

4. For high voltages, overhead-line construction is generally used with suspension-type insulators. Steel towers, called pylons, serve to carry the insulators, with each conductor suspended from the bottom of a group or string of insulator units.

Section C
Hydropower Technology Development

Hydroelectric power (Fig. 3-9) is the largest source of renewable electricity in the United States, producing about 6.3% of the nation's total electricity throughout the last decade. Even after a century of proven experience with this reliable renewable resource, significant opportunities still exist to expand the nation's hydropower resources through non-powered dams, water conveyance systems, pumped storage hydropower, and new site development. The Water Power Program supports the hydropower industry and complements existing investments through the development and deployment of new technologies and key components, as well as by identifying key opportunity areas through which hydropower generation can be enhanced.

Fig. 3-9 Hydroelectric power station

In addition to conventional hydropower, pumped-storage hydropower is an important piece of DOE's renewable energy portfolio because it acts as utility-scale grid storage technology. The Water Power Program can play an essential and catalytic role in demonstrating the benefits of pumped-storage hydropower as a part of our clean energy future—acting as a renewable form of grid stabilization and enabler for the high penetration of variable renewables (such as wind and solar). A U.S. Department of Energy 2015 report to Congress outlines key activities that can help accelerate pumped-storage development in the United States.

With more than 2,500 U.S. companies supporting the hydropower industry, adding additional hydropower generation will create a large and enduring economic benefit by revitalizing the domestic manufacturing and hydropower industry.

Learn more about the Water Power Program's work in the following areas of hydropower technology development:

- Low-Head Hydropower
- Materials and Manufacturing
- Hydropower Systems
- Hydropower Technology Accomplishments

Low-Head Hydropower

There is a significant opportunity across the country to add new hydropower generating capabilities at low-head sites (i.e., those that operate with a change in elevation ranging from 2 to 20 meters). These types of waterways are often present at existing non-powered dams, canals, and conduits across diverse areas of the United States. The Water Power Program is investing in innovative low-head hydropower technology R&D, such as Percheron Power's installation of the nation's first Archimedes Hydrodynamic Screw system. This project demonstrated that the low-head technology is simple, robust, and economical.

Materials and Manufacturing

The Water Power Program funds R&D to identify and test new materials and manufacturing techniques to improve the performance and lower the costs of hydropower. Program-funded research focuses on materials or coatings that reduce the life-cycle cost of turbine runners, draft tubes, and penstocks. R&D also focuses on identifying and testing ways to improve generator efficiency and reliability.

Hydropower Systems

The Water Power Program works to develop and test new technologies and techniques that can reduce operations and maintenance costs; increase unit availability and plant capacity factors; reduce risk through enhanced system reliability; and improve the quality—environmental performance attributes, as well as ancillary power benefits—of the energy produced. Areas of focus include water-use optimization, the application of advanced materials and manufacturing methods, and the assessment of the value of water power grid services. For example, existing hydropower facilities in the United States show signs of deterioration, and the data used to evaluate these facilities are scattered and outdated. The Water Power Program is working with partners to integrate and update information in order to understand the declines in electricity generation, capacity factors, and facility availability.

Technology Development Accomplishments

The program has numerous accomplishments in hydropower technology development. The projects described below highlight just a few of the program's new opportunities and recent successes in cost reductions, water-use optimization, and facility upgrades.

New opportunities for advanced hydropower R&D In 2011, after revamping its hydropower technology efforts, the Water Power Program released its first major solicitation for hydropower R&D in more than a decade. These projects aim to reduce costs of hydropower technologies and demonstrate the dynamic grid benefits of advanced hydropower and pumped storage technologies. For example, Natel Energy, in consultation with Alden Research Laboratory, designed, built, and commissioned a reliable powertrain for the Schneider Linear hydroEngine^{TM}. By reducing capital and maintenance costs, this powertrain enables the development of new low-head hydropower capacity—achieving levelized-cost-of-energy savings of around $2 per megawatt hour.

Optimizing hydropower systems for power and environment The Water Power Program sponsored a team of U.S. national laboratories to develop and demonstrate a suite of advanced, integrated analytical tools, known as the Water-Use Optimization Toolset (WUOT). WUOT assists managers and operators with operating their hydropower plants more efficiently, resulting in more energy and grid services from available water resources, thus enhancing the environmental benefits from improved hydropower operations and planning. WUOT includes tools for hydrologic forecasting, seasonal hydro-systems analysis, day-ahead scheduling, real-time operations, and environmental performance operations. The following locations are deploying WUOT for demonstration:

- The Oroville Complex on the Feather River in California.
- The upper Colorado River portion of the Colorado River Storage Project.
- The Conowingo Dam on the Susquehanna River in Maryland.

Revitalizing American infrastructure Sponsored through the American Recovery and Reinvestment Act of 2009, the Water Power Program completed three hydropower efficiency projects with overwhelming success—resulting in an increase of more than 3,000 megawatt-hours per year.

• The Los Alamos County Department of Public Utilities installed a low-flow turbine to its Abiquiu Hydroelectric Facility in New Mexico. The new turbine boosts overall facility output from 13.8 megawatts to 16.8 megawatts.

• The City of Boulder in Colorado completed a modernization project to its Boulder Canyon Hydroelectric Project by installing a new turbine/generator unit. The new unit resulted in a 30% increase in generation and an 18% - 48% increase in turbine efficiency.

• The City of Tacoma installed two Francis turbine/generator units to the Cushman Dam in Washington. The new units add approximately 3.6 megawatts of annual electrical generation.

(943 words)
Office of Energy Efficiency & Renewable Energy
https://www.energy.gov/eere/water/hydropower-technology-development

New Words and Expressions

ancillary [æn'siləri] a.	辅助的,附加的
catalytic [ˌkætə'litik] a.	促进性的;起催化作用的
coating n.	涂层,镀膜
commission [kə'miʃn] v.	调试;委托
complement ['kɔmplimənt] v.	补充,补足
conduit ['kɔndjuit] n.	管道
conveyance [kən'veiəns] n.	运输,输送
deterioration [diˌtiəriə'reiʃən] n.	恶化,变坏
hydrologic [ˌhaidrə'lɔdʒikəl] a.	水文的
megawatt ['megəwɔt] n.	兆瓦,百万瓦特
penstock n.	压力管道
portfolio [pɔːt'fəuliəu] n.	系列计划;文件夹
powertrain n.	动力系统,传动系统
revamp [ˌriː'væmp] v.	改变,修改;翻新
solicitation n.	(意见的)征求
draft tube	尾水管
levelized-cost-of-energy	能源平准化成本
low-flow turbine	低流量涡轮机
low-head hydropower	低水头水力发电
pumped-storage hydropower	抽水蓄能水力发电

turbine runner   涡轮机转轮
utility-scale grid   公用事业规模输电网

Notes

1. Archimedes Hydrodynamic Screw system   阿基米德水动力螺旋系统
2. DOE (U. S. Department of Energy)   美国能源部
3. R&D (research and development)   研发,研究与开发
4. the Schneider Linear hydroEngine   施耐德线性液压发动机
5. the Water Power Program   水力发电计划
6. the Water-Use Optimization Toolset (WUOT)   用水优化工具集

📖 Text Understanding

Ⅰ. Decide whether the following statements are true (T) or false (F) according to the passage.

1. Hydroelectric power in the United States produces about 6.3% of the nation's total electricity in the last decades.
2. The Water Power Program supports the hydropower industry in various ways.
3. Additional hydropower generation will create a large and enduring economic benefit.
4. Low-head sites refer to the locations where their elevations are between 2 to 20 meters.
5. The Schneider Linear hydroEngine is the joint effort of Natel Energy and Alden Research Laboratory.
6. A low-flow turbine in Abiquiu Hydroelectric Facility increases its facility output by 16.8 megawatts annually.

Ⅱ. Fill in the blanks with proper words according to the passage.

1. _____ is the largest source of renewable electricity in the United States.
2. _____ acts as utility-scale grid storage technology.
3. Low-head technology is _____, _____, and _____.
4. The Water Power Program achieved many successes in _____, _____, and _____.
5. A team of U.S. national laboratories develop and demonstrate a suite of advanced, integrated analytical tools, known as the _____.

III. Give brief answers to the following questions.

1. Which areas does the Water Power Program's work in hydropower technology development cover?

   _____

2. What is the purpose of the Water Power Program's funding R&D?

   _____

3. What tools does WUOT include?

   _____

Vocabulary Building

IV. Fill in the table below by giving the corresponding translation.

English	Chinese
	调试
	兆瓦
powertrain	
turbine runner	
	涂层
pumped-storage hydropower	

V. Fill in the blanks with the words from the passage. The first letter of the word is given.

1. Significant opportunities still exist to expand the nation's hydropower resources through non-powered dams, water c_____ systems, pumped storage hydropower, and new site development.

2. Under the framework of the a_____ services pricing, this dissertation analyzes the value compensating mechanism of the pumped-storage plant.

3. The Water Power Program plays an essential and c_____ role in demonstrating the benefits of pumped-storage hydropower.

4. Existing hydropower facilities in the United States show signs of d_____, and the data used to evaluate these facilities are scattered and outdated.

5. The department was r_____ to try to improve its performance.

✎ Translation Practice

Ⅵ. Translate the following paragraph into English.

水电站使用水轮机,水轮机所需要的水是从高于涡轮处落下或流过的,由此驱动轴做功。流向涡轮的水在压力下高速强行通过喷嘴或叶片;弯曲的叶片使喷射流改变方向并对涡轮上旋转的叶片施加一个作用力。叶片安装在轮和轴上,轴转动就会驱动与之直接相连的发动机。用过的水从涡轮流入较低地势的河中。

💬 Translation Skill Ⅲ

科技英语的特点

科技英语属于科技文体。科技文体是自然科学家和社会技术人员从事专业活动时使用的一种文体,如科学著作、学术论文、实验报告、产品说明书等。科技文体不以语言的艺术美为追求目标,而是讲求逻辑的条理清楚和叙述的准确严密。因此,科技英语有自己的语言、词汇和语法特点。试比较下面两个英语句子:

(1) The factory turns out 100,000 cars every year.
(2) The annual output of the factory is 100,000 cars.

例(1)和例(2)句意相同,但在用词和语法上有所差异。例(1)以动词短语作谓语,强调实际动作,且所用词汇均为日常常用词汇;例(2)以系表动词作谓语,强调对事实的客观陈述,句中的 annual 和 output 两词为比较正式的专业词汇。

我们通过表1对比日常英语和科技英语的文体差异。

日常英语和科技英语的文体差异　　　　　　表1

日　常　英　语	科　技　英　语
1. 通俗化:常用词汇用得多	1. 专业化:专业术语用得多
2. 多义性:一词多义,使用范围广	2. 单义性:语义相对单一,因专业而不同
3. 人称化:人称丰富,形式多样	3. 物称化:多用物称,以示客观
4. 多时性:描述生活,时态多样	4. 现时性:叙述事实,多用现在时
5. 主动态:句子倾向于主动语态	5. 被动态:句子倾向于被动语态
6. 简单性:单句、散句用得多	6. 复杂性:复杂句、复合句用得多
7. 口语化:口语用得多,随意灵活	7. 书面化:长句用得多,书卷气浓

下面,我们将结合实例分别从科技文体的词汇特点、动词的时态和语态特点以及名词化

和动词非谓语结构三个方面进行阐释。

1. 科技文体的词汇特点

(1) 大量使用专业术语

科技文章要求概念清楚，避免含糊不清和一词多义，因此使用较多的科技词汇。科技词汇来源主要有三类：

A. 来自英语中的日常词汇，但被赋予了新的词义。例如：

Work is the transfer of energy expressed as the product of a force and the distance through which its point of application moves in the direction of the force.

在上例中，work，energy，product，force 都是从日常词汇中借用来的物理学术语，在此句中，它们分别表"功、能量、乘积和力"之义。

B. 从希腊语或者拉丁语中派生而来。例如：

therm－热（希腊语）　　　　thesis－论文（希腊语）
astro－天空（拉丁语）　　　ampl－扩大（拉丁语）

C. 新造词汇。主要有以下几种构词方法：

a. 派生法（Deviation）。如表 2 所示。

由前缀后缀派生的新词　　　　　表 2

Prefix	Suffix	Examples
anti－相反		antibody, antipathy
geo－地球,土地		geography, geomagnetic
sub－下,低于		submarine, subcontinent
ultra－极端		ultrafilter, ultrasound
	-ology 学科	biology, anthropology
	-itis 炎症	arthritis, gastritis
	-meter 和丈量有关	gasometer, thermometer
	-ics 学科	statistics, physics, mechanics

b. 复合词（Compound Words）。例如：

work+shop → workshop　　　　lap+top → laptop
out+spread → outspread　　　　record+breaking → record-breaking

c. 缩略法（Abbreviation）。例如：

math ← mathematics　　　flu ← influenza　　　lab ← laboratory
CPU ← Central Processing Unit　　　GPS ← Global Positioning System

d. 混成法（Blending）。例如：

transceiver ← transmitter+receiver　　　hitech ← high+technology

(2) 倾向于使用比较正式的词汇

这类词汇不是专业术语，但由于其比较正式，在日常语言中使用较少。然而，其出现范围广泛，不仅出现在科技英语文体中，也出现在政治、经济、法律、语言等社会科学的文体中。例如：substantial，negligible，appreciable 等等。

2. 动词的时态和语态特点

(1) 时态

英语动词共有十六个时态，其中科技英语文体最常用的是一般现在时、一般过去时、一般将来时和现在完成时。

A. 一般现在时：

a. 一般叙述过程：Power engineering deals with the generation, transmission, and distribution of electricity.

b. 叙述客观事实或科学定理：The nucleus consists of two different kinds of particles, namely protons and neutrons.

c. 通常或习惯发生的行为：Alternating current is usually supplied to people's houses at 50 cycles per second.

B. 一般过去时：More than two thousand years ago the Greek philosopher Thales observed that when a piece of amber, a hardened gum from trees, was rubbed with a material like wool or fur, it attracted certain other kinds of material.

C. 一般将来时：In time, many things now unknown will become known.

D. 现在完成时：Electronic refinements have greatly extended the uses and capabilities of some of the older electrical devices.

(2) 语态

被动语态在科技文章中使用频繁，这主要有两个原因：第一，科技文章重在描写行为或状态本身，因此行为或状态的主体就显得不那么重要，他们要么没有必要指出，要么根本无法指出。虽然我们在科技文章中仍能看到带有 by 短语的句子，但其数量不多，且 by 之后的短语常常不是被动行为的主体，而是行为的方式或工具；第二，频繁使用被动语态有利于向后扩展句子，而不至于使句子头重脚轻。科技英语中被动语态的翻译方法将在后文第七章的翻译技巧中详细论述，故此不再赘述。

3. 名词化和动词非谓语结构

(1) 名词化

名词化是科技英语文体中非常普遍的现象，这是因为科技文体要求行文简洁、表达客观、内容确切、信息量大、强调存在的事实等。所谓名词化，就是把动词变成有动作含义的名词，如果是动词短语或句子，则把动词短语或句子变成名词短语。例如：

air moves → the motion of air

to apply force → the application of force

to translate Chinese into English → the translation of Chinese into English

名词化把大量原来的动词短语或句子变成了名词短语,原来的动作被转化成了事物。这样可以使句子变得比较简练,但变成名词短语以后,句子可容纳的信息量增多,表示的思想变得复杂,因此有可能生成更复杂的句子。所以说,名词化既是句子的简化手段,又是句子的复杂化手段。

(2) 动词非谓语结构

科技文章要求行文简练、结构紧凑,而英语每个简单句中只能有一个谓语动词,如果要同时描述几个动作,需要选取其中的主要动作当谓语,其余动作用非谓语动词结构表示,这样才符合英语语法规则。因此,在科技文章中,常用分词短语(或独立主格)代替各类从句。例如:

A direct current is a current which always flows in the same direction.

→ A direct current is a current always flowing in the same direction.

Vibrating objects produce sound waves and each vibration produces one sound wave.

→ Vibrating objects produce sound waves, each vibration producing one sound wave.

科技英语文体特点的总括让我们了解了科技英语文章的概貌,我们将在此基础上结合其文体特点分别从词语、句法和语篇角度探讨具体的科技文体英汉翻译技巧。

Unit 4

Electromagnetism

Before-Class Reading

Answer the following questions before reading the passage:

1. Do you know from which direction the invisible magnetic lines of force start in a magnet?
2. Can you give some examples of electrical devices making use of electromagnetism?

Section A
Introduction to Electromagnetism

In addition to static electricity, the ancient Greeks observed another natural phenomenon. They discovered a kind of iron ore which possessed the ability to attract or repel certain other kinds of material, especially iron. This property is called magnetism, and an object that possesses it is called a magnet. Iron containing a small amount of carbon can be made into a magnet by placing it in contact with another magnet; it is then said to have been magnetized.

When two bar magnets are brought together, one end of the first magnet attracts one end of the second magnet, but repels the opposite end of that magnet. If the direction of one magnet is reversed, the attraction and repulsion are also reversed. The two ends of a magnet are called north and south poles, shortened forms for "north-seeking" and "south-seeking" poles, because when magnet is allowed to pivot freely, one end always points toward the north and the other end toward the south. This is because the earth itself is a giant magnet with magnetic poles near the geographic North and South Poles.

A magnet does not simply exert a force of attraction in a straight line from one pole to the other. Rather, it establishes a magnetic field, in which the direction and strength of the force are indicated by lines of force. These extend out from each pole of the magnet and meet to form an oval-shaped arc (Fig. 4-1). Lines of force go out of the north pole and back into the south pole. Where the magnetic field is the strongest, there are proportionately more lines of force.

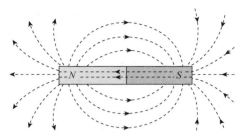

Fig. 4-1　An oval-shaped arc of magnetic field

There is an obvious similarity between the phenomena of electricity and magnetism. We have already noted that like electric charges repel and unlike charges attract. With magnets, like poles repel and unlike poles attract.

The relationship between electricity and magnetism was discovered accidentally by a Danish scientist, Hans Oersted, in 1819. He had left a compass on a table where he was experimenting with an electric current. A compass is a navigational device with a magnetized needle which points to the earth's north and south magnetic poles. Oersted observed that the needle moved whenever the current was turned on, and concluded that electric current possessed the property of magnetism. Oersted had discovered the phenomenon of electromagnetism, a discovery with momentous consequences. It has led to the development of many devices which make use of electromagnetism, including electric motors, generators, and transformers. Without these devices, electricity could never have become a major source of power.

Other scientists experimenting with electromagnetism found that the magnetic effect of an electric current could be strengthened by sending the current through a coil—a wire conductor twisted into a spiral shape. A greater number of turns of wire in the coil strengthen the magnetism, as does a stronger electric current. Wrapping the coil around a core of iron further increases the magnetism, because the iron itself becomes magnetized. All these discoveries led the way to converting electromagnet into motion. The great advantage of this energy conversion is that devices based on electromagnetism can be controlled simply by switching the current on or off.

Fig. 4-2　An electric motor

One such device is an electric motor (Fig. 4-2), in which a bar known as an armature (Fig. 4-3) is placed between the two arms of a horseshoe magnet. Magnetic poles are induced in the armature by sending a current through a coil wrapped around it. Magnetic force then causes it to move in the direction in which the unlike poles attract each other. Of course, if the poles of the armature reached a position directly opposite to the unlike poles of the horseshoe magnet, the armature would become locked and no further motion would be possible. It is necessary, therefore, to prevent the unlike poles from becoming aligned. In a direct current electric motor, this is done by a device called a commutator (Fig. 4-4), which reverses the electron flow, changing the magnetic poles of the armature at each half-turn. This causes the armature to move on to the new poles of attraction, completing a full rotation, or "cycle".

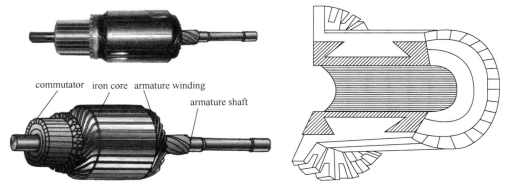

Fig. 4-3　An armature　　　　　　　　Fig. 4-4　A commutator

In an alternating current electric motor, the stationary portion is called a stator (Fig. 4-5) and the rotating portion, a rotor (Fig. 4-6). With alternating current, the flow of electrons reverses at rapid intervals. The intervals are timed to change the stator and rotor poles simultaneously so that the rotor continues to move in a circle, or cycle. In the United States, alternating current electric motors are timed to complete 60 cycles per second, but in Europe they are timed for 50 cycles per second.

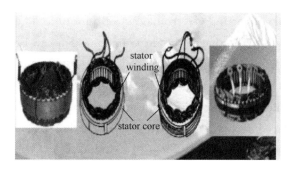
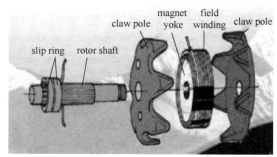

Fig. 4-5　A stator　　　　　　　　Fig. 4-6　A rotor

The property of electromagnetism is also vital to the production of electrical power. Electricity from chemical action—the cell or battery—is suitable only for special and limited uses. Generators based on electromagnetism, however, produce sufficient cheap electricity to supply most of the world's needs.

Not long after the discovery that a magnetic field could be created by an electric current, the English scientist Michael Faraday discovered that the reverse was also true—a current could be created by a magnetic field. When a closed loop of wire moves through a magnetic field, an electromotive force (e.m.f.) is created. This causes a current of electrons to flow through the wire and is the basis for generating electricity. Because it is electromotive force that is produced, the power of generators is described in terms of volts, the units of measurement for e.m.f. To generate e.m.f., the wire must cut the lines of force in the magnetic field; if the wire moves parallel to them, no e.m.f. is produced. Also, the faster the wires are made to move, the greater the production of e.m.f. For this reason, a rotary (circular) motion is used in a generator rather than a reciprocating (up-and-down or back-and-forth) motion—it is much easier to maintain a rotary motion at high speeds.

A generator contains a stator, a stationary magnet, with a rotor placed between its north and south poles. As the rotor turns, the wires in it cut the lines of force in the magnetic field of the stator. With each half-turn, the current flow is reversed. This is what produces alternating current, in which the electron flow rapidly changes direction over and over again.

The rotor is turned by a turbine, a machine with huge blades which are generally moved by water or steam. Steam for an electric power plant can be obtained from the heat supplied by burning coal or oil, or by nuclear fission.

The enormous size of modern generators and the speed with which the rotor can turn mean that electric power of very high voltage can be produced. As much as half a million volts can be transmitted over high-voltage lines to substations. There the voltage is reduced to strengths that can be used in factories or homes. In the United States the customary voltage for household use has become 110-120V, since it was discovered that higher voltages could cause fatal accidents. 220V is still supplied for some heavy-duty uses, however, such as for operating an electric stove.

Transformers (Fig. 4-7) are the devices that increase or decrease the amount of voltage. Like generators, they also depend on electromagnetic effects. A transformer basically consists of two coils of wire wrapped around iron cores. Current is supplied to a transformer through the primary coil and is taken from the secondary coil. When an alternating current passes through the primary coil, the constant reversal of electron flow produces a changing magnetic field

Fig. 4-7 Transformers

that creates a current in the secondary coil.

When the primary coil has more turns than the secondary coil, the secondary voltage is decreased; this is a step-down transformer. When the secondary coil has more turns than the primary, the secondary voltage is increased; this is a step-up transformer. Step-up transformers are used at a power plant to increase the voltage for transmission. As we have pointed out, the higher the voltage in transmission, the less energy loss takes place. At substations that distribute electricity for use, and in all buildings or houses supplied with electricity, there are step-down transformers to change the power to voltages suitable for industrial or domestic use.

(1406 words)

Quoted from *English for Electrical and Electronic Science and Technology*

by Bu Yukun

New Words and Expressions

align [ə'lain] v.	校准,排整齐
armature ['ɑ:mətʃə] n.	电枢(电机的部件)
commutator ['kɔmjuteitə] n.	换向器
heavy-duty a.	重型的,重负荷的;耐用的
navigational [ˌnævi'geiʃənl] a.	航行的,航运的,导航的
pivot ['pivət] n. & v.	支点,枢轴 & 在枢轴上旋转
proportionately [prə'pɔ:ʃənətli] adv.	成比例地
repulsion [ri'pʌlʃn] n.	排斥,反感
spiral ['spairəl] a.	螺旋形的
stator ['steitə] n.	定子
bar magnet	条形磁铁,磁棒
closed loop	闭合回路
horseshoe magnet	蹄(U)形磁铁
iron ore	铁矿
line of force	力线
oval-shaped arc	椭圆形弧线
primary coil	初级线圈
secondary coil	次级线圈
step-down transformer	降压变压器
step-up transformer	升压变压器

Electrical Engineering **PART**

Exercises

📖 Text Understanding

I. Decide whether the following statements are true (T) or false (F) according to the passage.

1. The earth is a giant magnet with magnetic poles near the geographic North and South Poles.
2. The invisible magnetic lines of force are all straight lines.
3. The invisible magnetic lines of force start from the south pole and return to the north pole.
4. Electric motors, generators, and transformers are examples of the application of electromagnetism.
5. If the poles of the armature reached a position directly opposite to the unlike poles of the horseshoe magnet, the armature would become locked and no further motion would be possible.
6. The unit of measurement for e.m.f. is voltage.
7. In the United States the customary voltage for household use is 110V.
8. In power transmission, extremely high voltage is applied in that the higher the voltage, the less energy loss takes place.

II. Fill in the blanks with proper words according to the passage.

1. Where the magnetic field is the strongest, there are proportionately _____.
2. In 1819, a Danish scientist, _____ accidentally discovered the relationship between electricity and magnetism.
3. To prevent the unlike poles from becoming aligned, a device called a _____ is used.
4. In an alternating current electric motor, the stationary portion is called a _____ and the rotating portion is called a _____.
5. A _____ motion is circular and a _____ motion is up-and-down or back-and-forth.
6. _____ are the devices that increase or decrease the amount of voltage.

III. Give brief answers to the following questions.

1. What is the obvious similarity between electricity and magnetism?

2. Why is it necessary to prevent the unlike poles from becoming aligned?

3. What is the difference between a step-up transformer and a step-down transformer? And where are they used respectively?

Vocabulary Building

Ⅳ. Fill in the table below by giving the corresponding translation.

English	Chinese
	初级线圈
pivot	
	闭合回路
bar magnet	
	电枢
step-up transformer	

Translation Practice

Ⅴ. Translate the following paragraph into English.

大约在公元1800年,人们发现,当指南针靠近带电的导线时会受其影响。一位叫作Hans Christian Oersted的丹麦工程师于1820年对此现象进行了研究。他得出结论,当电流通过导线时会在导线周围产生一个磁场,该磁场呈环形状。既然一条导线中的电流能产生一个磁场,那么多条导线带有相同的电流是不是能产生更强的磁场呢?事实证明确实如此。导线数量的增加可以通过线圈的形式来实现。

In-Class Reading

Section B
Direct Current Generators and Motors

The conversion of mechanical power into electrical power, and conversely of electrical power into mechanical power, is one of the most important problems of electrical engineering. These conversions are made by means of electric machines. When the machine is driven by mechanical torque, it converts mechanical energy into electrical energy and is called a generator; when it is driven electrically, by passing a current through it from some external source, it converts electrical energy into mechanical energy and is called a motor. In the case of d.c. machines, generators and motors have an identical structure. In fact, the same machine may be used as a generator or as a motor, as is shown in Fig. 4-2.

D.c. energy conversion machines use two basic principles:

1. When a conductor is moved through a magnetic field in such a way as to cut the magnetic lines, an e.m.f. is generated in the conductor;

2. When a current flows along a conductor in a magnetic field, a mechanical force is produced that tends to move the conductor.

Fleming's right-hand rule (Fig. 4-8) gives us a rapid way of relating the directions of the flux (field), motion and e.m.f. (current) when we wish to apply these principles in practice to electric machines. A d.c. generator consists of a field structure—a series of alternate north and south magnetic poles spaced around a circular periphery. The poles, two or any other even number, are mounted on the frame or yoke. This has two functions: it

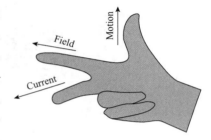

Fig. 4-8 Fleming's right-hand rule

is a portion of the magnetic circuit, and it can act as a mechanical support for the machine as a whole. The armature consists of a cylindrical core of iron laminations insulated from one another and mounted on a shaft so that it may rotate in the magnetic field produced by the poles. In slots on the surface of the armature are paced copper conductors, which are insulated from the core with treated fabric (paper or bonded mica flakes) and held in the slots by insulating wedges or wire bands. The poles near the armature have enlarged portions that are called the pole shoes, and the areas facing the armature are referred to as the pole faces. The spaces between the pole faces and the armature are the air gaps. The faces of the poles are practically coaxial with the armature and the lengths of the air gaps in the magnetic circuit are relatively small, thus making the magnetic reluctance relatively small.

The magnetomotive force necessary to establish the flux in the magnetic circuit is obtained by means of the field coils wound on the poles. The exciting current for the field coils may be supplied in various ways. When a generator supplies its own exciting current, it is said to be self-excited; when the exciting current is supplied from an external source, the machine is said to be separately excited.

In a d.c. generator, armature conductors are moved through the magnetic field in such a manner that e.m.f.s are induced in them. These conductors are arranged in a winding in such a way that certain groups of conductors are in series between two points of the winding, care being taken when the connections are made that no induced e.m.f.s are in opposition. By means of the commutator and the brushes, contact between the armature winding and the external circuit is made (Fig. 4-9). In other words, we may say that in each single conductor we have an e.m.f. that varies while the conductor is moving through the air gap. In fact the e.m.f. will be greatest when the conductor is under a pole, and zero when the conductor is central between two poles. Moreover, the e.m.f.s generated in all the armature conductors under a north pole are obviously all in the same direction, while those generated in conductors under a south pole are in the opposite direction. If, on the other hand, we consider a fixed position in the air gap, we always have the same e.m.f. in all conductors in that position.

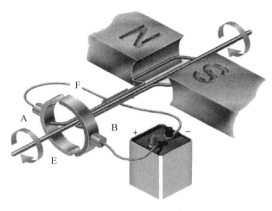

Fig. 4-9 The working principle of the commutator

The brushes are fixed with respect to the rotating armature, hence the e.m.f. between them is constant and they transmit a constant voltage to the external circuit. The commutator is composed of a number of copper segments insulated from each other and from the spider and clamping flanges by bonded mica flakes. Brushes are generally made of graphitized carbon. They fit into metal brush-holders and are held against the commutator by spring pressure.

When a commutator segment passes under a brush, the current in the armature conductor connected with it must reverse from a given value in one direction to the same value in the opposite direction. This is known as commutation. Since the armature coils lie in magnetic material, they are inductive, and the reversal of the current during commutation develops voltages of self and mutual induction in the commutator segments that are shorted by the brushes. If these voltages are large, the result is that current continues to flow directly from the segment of commutation to the brush even after contact has been broken. This flow of current forms a spark that can burn the surface of the commutator. It is customary to reduce sparking by inducing a voltage equal and opposite to that caused by the change in current. Shifting the brushes will do this, but a different angle of shift is needed for each load condition. The use of an interpole or commutating pole whose winding carries the same current as the armature is more satisfactory as a method of inducing an opposite voltage.

As we have said, d.c. motors have the same structure as d.c. generators Fig. 4-10 is the panoramic view of a generator or a motor. In a motor, current flows through the armature conductors from an external source. These conductors carrying current produce a magnetic field that in the presence of the stator magnetic field produces a torque which tends to turn the armature. The field coils and armature windings used in motors are the same as in generators and the problem of commutation is also similar.

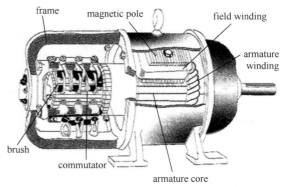

Fig. 4-10 The panoramic view of a generator/motor

It is very important to know the efficiency of generators and motors. This can be calculated by output/(output+losses) for the former and by (input-losses)/input for the latter. In both cases the principal losses are: mechanical loss, iron loss, copper loss in the field and armature coils and commutator friction and resistance loss.

(1094 words)

Quoted from *English for Electrical and Electronic Science and Technology* by Bu Yukun

New Words and Expressions

alternate ['ɔːltəneit] a. 交替的
brush n. 电刷
brush-holder n. 电刷支架
commutation [ˌkɔmjuˈteiʃn] n. 换向
flux [flʌks] n. 磁通量
graphitized [ˈɡræfəˌtaizd] a. 石墨化的
interpole [ˈintəpəul] n. 换向极
lamination [ˌlæmiˈneiʃən] n. 迭片结构,层压
panoramic [ˌpænəˈræmik] a. 全景的
self-excited a. 自励磁的,自激的
slot [slɔt] n. 槽
spider n. 支架,三脚架
torque [tɔːk] v. 扭矩,转矩
wedge [wedʒ] n. 楔子,楔形物

winding [waindiŋ] *n.*	绕组
yoke [jəuk] *n.*	轭铁
air gap	气隙
bonded mica flake	屏蔽云母片
clamping flange	钳式法兰
commutating pole	整流极
copper segment	换向器铜片
even number	偶数
exciting current	励磁电流
field coil	励磁线圈
in series	串联
magnetic reluctance	磁阻
magnetomotive force	磁动势,磁通势
mutual induction	互感
pole face	极面
pole shoe	极靴,极端
self induction	自感,自感现象
separately excited	他励磁的,分激的
spring pressure	弹力
treated fabric	处理过的纤维
wire band	线圈

Notes

1. Fleming's right-hand rule   弗莱明右手定则

📖 Text Understanding

Ⅰ. **Choose the best answer according to the passage.**

1. A _____ converts electrical energy into mechanical energy.

   A. generator   B. motor   C. distributor   D. rectifier

2. Fleming's right-hand rule takes the following three factors into consideration except _____.

   A. flux   B. motion

C. direction   D. electromotive force

3. The number of north and south poles in a field structure of a d.c. generator cannot be _____.
   A. nine   B. six   C. eight   D. twelve

4. In a d.c. generator, certain groups of armature conductors are connected _____.
   A. at random   B. crosswise
   C. in parallel   D. in series

5. The efficiency of motors can be calculated as _____.
   A. output/(output+losses)   B. (input-losses)/input
   C. output/(output-losses)   D. (input+losses)/input

II. Fill in the blanks with proper words according to the passage.

1. In the case of d.c. machines, generators and motors have an _____ structure.
2. The armature consists of _____ insulated from one another and mounted on a shaft.
3. The poles near the armature have enlarged portions that are called _____, and the areas facing the armature are referred to as _____.
4. In a d.c. generator, the e.m.f. will be _____ when the armature conductor is under a pole, and _____ when the conductor is central between two poles.
5. The use of an interpole or commutating pole is more satisfactory than _____ as a method of inducing an opposite voltage.

III. Give brief answers to the following questions.

1. Why are poles mounted on the frame or yoke?

2. Why are pole faces practically coaxial with the armature and the lengths of the air gaps in the magnetic circuit are relatively small?

3. What is a self-excited generator and what is a separately excited generator?

4. What is commutation?

5. How to solve the sparking problem?

Vocabulary Building

IV. Fill in the table below by giving the corresponding translation.

English	Chinese
lamination	
interpole	
	电刷支架
	励磁电流
in series	
spring pressure	
	气隙
	互感

V. Match the items listed in the following two columns.

1. flux
2. self-excited
3. torque
4. winding
5. field coil
6. wedge

a. a generator supplies its own exciting current
b. the electric coil around a field magnet that produces the magneto-motive force to set up the flux in an electric machine
c. an electrical conductor such as a wire in the shape of a coil, spiral or helix
d. the rate of flow of energy or particles across a given surface
e. twisting force causing rotation in machinery
f. any shape that is triangular in cross section

Translation Practice

VI. Translate the following sentences from the passage into Chinese.

1. When a conductor is moved through a magnetic field in such a way as to cut the magnetic lines, an e.m.f. is generated in the conductor.

2. When a current flows along a conductor in a magnetic field, a mechanical force is produced that tends to move the conductor.

Electrical Engineering **PART 2**

3. A d.c. generator consists of a field structure—a series of alternate north and south magnetic poles spaced around a circular periphery.

_____
_____
_____

4. The commutator is composed of a number of copper segments insulated from each other and from the spider and clamping flanges by bonded mica flakes.

_____
_____
_____

5. Brushes are generally made of graphitized carbon. They fit into metal brush-holders and are held against the commutator by spring pressure.

_____
_____
_____

Section C
Transformers

When we introduced the subject of alternating current, we said that the very great use made of a.c. supply systems is due to the ease with which an a.c. supply can be stepped up or down in voltage. To do this we use a static transformer (Fig. 4-11). This device transforms low-voltage energy into high-voltage energy or vice versa.

Fig. 4-11 Transformers in a substation

Transformers are also required for the operation of most luminous discharge tubes and for bell systems in the home. They are found in radio and television equipment, with several different types often used in the same circuit. Telephone systems use large numbers

81

of transformers (Fig. 4-12), as does X-ray equipment. As a result, we might say that where electricity is used, transformers are used too.

Fig. 4-12 Miniature transformers

Essentially a transformer consists of two independent electric circuits linked with a common magnetic circuit. In particular, it usually consists of two coils or windings of insulated copper wire around a laminated iron core. An alternating magnetic flux is produced in this iron core when an alternating current passes through the insulated conductors. In operation, an a.c. supply is connected to one winding which thereby becomes the primary winding; the other winding, to which the load is connected, is called the secondary winding.

The magnetic flux in the magnetic circuit formed by the iron core links the turns of the primary and secondary windings. If this flux is the same for each of the two windings, the e.m.f. induced per turn must be the same in each winding, so that the total induced e.m.f. in each winding will be proportional to the number of turns on that winding.

When the primary winding is connected to an a.c. supply with the secondary winding open, we say that this is a no-load operation. In this case the primary winding behaves like a reactance coil and takes an exciting current from the line to magnetize the iron core. This exciting current can be neglected in comparison with the value of the full-load current of the transformer.

The continuously changing flux in the iron core produces power loss. Loss in the iron is due partly to hysteresis loss and partly to eddy-current loss. Hysteresis loss is the power required for the continual reversals of the molecular magnets of which the iron is composed; eddy currents are due to the e.m.f.s induced in the core material as an effect of the varying flux density. In transformers generally, special grades of steel alloyed with silicon are used to give low eddy-current and hysteresis losses.

Because the primary and secondary windings do not occupy exactly the same space, some of the flux that links one winding does not link the other winding so that the e.m.f.s per turn of the two windings are not exactly the same. This flux is called the leakage flux.

To round off our schematic description of transformers, mention must be made of the iron loss in the iron core and the copper loss in the windings. These cause a temperature rise in the transformer; and unless arrangements are made to dissipate the heat, the winding insulation will begin to deteriorate. The means used to obtain cooling enable us to classify transformers into the following main types:

1. dry natural cooled;
2. dry forced-air cooled;
3. oil-immersed self-cooled;
4. oil-immersed forced-air cooled;
5. oil-immersed water cooled.

(568 words)

From *Fundamentals of Electrical Engineering*

by Johnson D.

New Words and Expressions

alloy ['æləi] n.	合金
deteriorate [di'tiəriəreit] v.	变坏,恶化
dissipate ['disipeit] v.	消散,驱散
luminous ['lu:minəs] a.	发光的,发亮的
reactance [ri'æktəns] n.	电抗
schematic [ski:'mætik] a.	略图的,图解的
silicon ['silikən] n.	硅
be proportional to	与……成正比
discharge tube	放电管
eddy-current loss	涡流损耗
full-load current	满负载电流,全负荷电流
hysteresis [histə'ri:sis] loss	磁滞损耗
laminated iron core	迭片式铁芯
leakage flux	渗漏通量
molecular magnet	分子磁铁
no-load operation	空载运行
round off	完成,结束

Exercises

📖 Text Understanding

Ⅰ. **Decide whether the following statements are true (T) or false (F) according to the passage.**

1. Transformers are of different sizes, from large transformers used in a substation to smaller ones used in telephone systems.
2. The primary winding always has fewer turns of coil than the secondary winding.
3. If the flux is the same for the primary and secondary windings, the e.m.f.s induced per turn must be the same in each winding.
4. Hysteresis loss is due to the e.m.f.s induced in the core material as an effect of the varying flux density.

Ⅱ. **Fill in the blanks with proper words according to the passage.**

1. A transformer usually consists of two independent _____ linked with a common _____.
2. The primary winding of a transformer is connected to _____, the second winding to _____.
3. The total induced e.m.f. in each winding will be _____ to the number of turns on that winding.
4. The two kinds of power losses in the iron core are _____ and _____.

Ⅲ. **Give brief answers to the following questions.**

1. Illustrate some examples to testify that "where electricity is used, transformers are used too".
2. What is a no-load operation?
3. How to solve the hysteresis and eddy-current losses problem in transformers?
4. What are the five types of transformers?

✎**Translation Practice**

Ⅳ. Translate the following paragraph into English.

启动汽车时,电启动电机中的电流生成磁场使电机轴旋转并驱动活塞压缩汽油与空气的爆炸性混合物。点火的火星来自放电,而放电又会生成短时电流。该现象看起来很复杂,实际上却来源于一些基本电磁现象的定律,其中最重要的定律之一是库仑(Coulomb)定律,它描述了带电体之间的电作用力。

Ⅴ. Translate the following paragraph into Chinese.

Everyday modern life is pervaded by electromagnetic phenomena. When a light bulb is switched on, a current flows through a thin filament in the bulb. The current heats the filament to such a high temperature that it glows, illuminating its surroundings. Electric clocks and connections link simple devices of this kind into complex systems such as traffic lights that are timed and synchronized with the speed of vehicular（车的）flow. Radio and television sets receive information carried by electromagnetic waves traveling through space at the speed of light.

💬 Translation Skill Ⅳ

词语的处理(1)—— 词义的引申

在科技英语汉译过程中,有时会遇到在词典上找不到与原文对等的对应词,或虽然有对应词,但不符合汉语的表达习惯,或出于修辞等原因,词的字面意义与表达意义不一致等情况。遇到上述情况时,我们需要根据语气、逻辑关系、搭配习惯或该词语在科技语篇中的专业含义等方面综合运用引申这一翻译技巧。词义的引申处理技巧主要有四类。

1. 专业化引申

科技英语文章中所使用的词汇有些来源于日常生活用语,但在科技语篇中被赋予了不同于日常语境的专业化语义。专业化引申的目的就是要使译文中涉及科技术语概念的词语符合其专业语言规范及习惯。例如：

(1) But because the <u>pump</u> is driven by the engine, the injection pressure rises and falls along with the engine <u>speed</u>.

(2) This should highlight the generally existing requirement of <u>high switching frequency</u> or <u>high power density</u>.

(3) The Lupo is billed as the world's first three-liter car—the number referring not to be the size of its engine, but to its <u>meagre thirst</u>.

例(1)的语境是汽车工程领域,结合该语境的专业术语表达,pump 由"泵"专业化引申为"输油泵",speed 由日常用语中的"速度"专业化引申为"转速"。例(2)的语境是电气工程领域,形容词 high 搭配不同的中心词 switching frequency(开关频率)和 power density(功率密度),分别被译为"高"和"大"。例(3)的语境同样是汽车工程领域,例中的 meagre 是法语词被借用到英语中,其日常语义为"瘦的、贫乏的",但如果此处采取直译则显得译文别扭、生硬,只能结合上下文语境运用专业化引申得出适当语义。以下为例(1)~例(3)的参考译文:

(1) 但因为<u>输油泵</u>由发动机驱动,所以喷油压力随着发动机的<u>转速</u>起伏。

(2) 应该强调普遍存在的<u>高</u>开关频率或<u>大</u>功率密度的要求。

(3) 路波被宣传为世界上第一辆 3 公升汽车,这个数字并不是指它的发动机尺寸,而是指它<u>极低</u>的耗油量。

2. 抽象化引申

科技英语文章中有些词语的字面意思比较具体或形象,如果直译成汉语会显得牵强,不符合汉语的表达习惯,使人感到费解。在这种情况下,就应该用含义较为概括或抽象的词语对英文所表达的词义加以引申。例如:

(4) Alloy belongs to a <u>half-way house</u> between mixtures and compounds.

(5) <u>On the downside</u>, the performance of a computer vision application is very dependent on the illumination condition.

(6) These gases <u>trap</u> the sun's heat whereas sulfur dioxide cools the atmosphere.

在例(4)中,half-way 的常规语义为"中途的、半路的",语义容易理解,但中心词 house(房子、旅店)和 mixtures(混合物)及 compounds(化合物)看似毫无关联,因此需要将 house 这个具体名词抽象化为上一层级属性——物质,在"物质"这一属性中,三个名词具有了共通性。例(5)中 downside 的日常语义为"底部",此处需抽象为"缺点、不足、负面"等语义进行理解。例(6)中的 trap 为动词,常规语义为"设陷阱捕捉"或"使陷入困境",但此语义与其宾语的逻辑关系不匹配,因此需将具体的动作抽象引申为"留住"。例(4)~例(6)的参考译文如下:

(4) 合金是介于混合物和化合物之间的<u>中间物质</u>。

(5) <u>缺点</u>是计算机视觉应用性能非常依赖照明条件。

(6) 这些气体<u>留住</u>了阳光中的热量,而二氧化硫使空气冷却。

3. 具体化引申

具体化引申是抽象化引申的反其道而行,需要把英文语句中含义较为概括、抽象、笼统的词语引申为意义较为具体、形象的词语,避免造成译文概念不清、不符合汉语语义逻辑或表达习惯的情况。例如:

(7) The <u>clever thing</u> about common-rail technology is that, by improving combustion, it dramatically reduces the level of both emission and noise.

(8) Public opinion is demanding more and more urgently that <u>something</u> must be done

about noise.

(9) The study of the brain is one of the <u>last frontiers</u> of human knowledge and of much more immediate importance than understanding the infinity of space or mystery of the atom.

在例(7)中,用 clever thing 的常规语义"聪明的事情"来指代共轨技术不符合中文的表达习惯,因此需要将较为抽象的常规语义"聪明的事情"具体化引申为"优势、优点",从而使译文清晰流畅。同理,例(8)中的 something 也应该根据上下文语境具体化、形象化,因此具体化引申为"某些措施"更加恰当。在例(9)中,如果将 last frontiers 直译为"最后的边疆"与上下文语境完全不匹配,因此结合脑科学、宇宙、原子研究等语境将其形象化引申为"最新领域"。以下为例(7)~例(9)的参考译文:

(7) 共轨技术的<u>优势</u>在于,通过改善燃烧,极大地降低了尾气排放和噪音水平。

(8) 公众舆论越来越强烈地要求为消除噪音采取<u>某些措施</u>。

(9) 对大脑的研究是人类知识的<u>最新领域</u>之一,它比研究无垠的宇宙和神秘的原子更迫切、更重要。

4. 修辞性引申

修辞性引申是指将英文原文语句中某个形象替换成译文汉语中读者所熟悉的形象,从而传达出英文原文的语用目的,由此使原文形象以间接的方式予以保留,使译文和原文有异曲同工之妙。例如:

(10) Recovery processing kills two <u>birds</u> with one <u>stone</u>.

(11) The test has fortified confidence that men should not be <u>chained</u> to the <u>wheel</u> of technology.

在例(10)中,kill two birds with one stone 是习惯用语,如果直译为"一石杀二鸟"虽保留了原文的形象,但汉语译文生涩,汉语读者需经过一番推敲才能体会其中的比喻义,故在传达喻义方面欠妥;通过修辞性引申将其译为"一箭双雕"既能间接保留英语原文形象,又能便于汉语读者的理解,应是接受度较高的译文。例(11)中 be chained to the wheel of 如果直译为"被拴在……车轮上"则显得翻译痕迹明显,用"让……牵着鼻子走"这一汉语惯用比喻形象替代英语原文形象,在保留原文形象的同时较好地完成了语用翻译。例(10)、例(11)的参考译文如下:

(10) 回收加工是一箭双雕。

(11) 该实验坚定了这一信念:人类不应当<u>让</u>技术<u>牵着鼻子走</u>。

综上所述,由于英汉文化和语言习惯差异,科技英语汉译时对词义进行引申处理是不可避免的常见现象。至于哪些词语采取哪种引申技巧,在很大程度上取决于汉语遣词造句的习惯。因此,在对英语词义引申处理时,应在不违背原文的前提下,尽可能使译文符合汉语的语言习惯,而不应拘泥于原文的语言形式。

Unit 5

Electrical Engineering Applications

Before-Class Reading

Answer the following question before reading the passage:

1. Do you like the idea of doing house chores by robots? Read the following passage and get familiar with the newly-emerged smart house.

Section A
Exploring Home Automation and Domotics

A smart house is a home that has highly advanced, automated systems to control and monitor any function of a house: lighting, temperature control, multi-media, security, window and door operations, air quality, or any other task of necessity or comfort performed by a home's resident. With the rise of wireless computerization, remote-controlled devices are becoming smart just-in-time. Today, it's possible to pin a programmed chip onto any occupant and have systems adjusted as a person passes by and through a smart house.

Is It Really Smart?

A smart home appears "intelligent" because its computer systems can monitor so many aspects of daily living. For example, the refrigerator may be able to inventory its contents, suggest menus and shopping lists, recommend healthy alternatives, and even routinely order groceries. The smart home systems might even ensure a continuously cleaned cat litter box or a house plant that is forever watered.

The idea of a smart home may sound like something out of Hollywood (Fig. 5-1). In fact, a 1999 Disney movie titled *Smart House* presents the comical antics of an American family that wins a "house of the future" with an android maid who causes havoc. Other films show science fiction visions of smart home technology that seems improbable.

However, smart home technology is real, and it's becoming increasingly sophisticated. Coded signals are sent through the home's wiring (or sent wirelessly) to switches and outlets that are programmed to operate appliances and electronic devices in every part of the house. Home automation can be especially useful for the elderly people with physical or cognitive impairments, and disabled persons who wish to live independently. Home technology is the toy of the super-wealthy, like Bill and Melinda Gates' home in Washington State, called *Xanadu* 2.0. The Gates' house is so high-tech that it allows visitors to choose the mood music for each room they visit.

Fig. 5-1　An android maid cleaning the house in a Hollywood movie

Open Standards

Think of your house like it's one big computer. If you ever opened up the "box" or CPU of your home computer, you'll find tiny wires and connectors, switches and whirling discs. To make it all work, you have to have an input device (like a mouse or a keyboard), but even more importantly, each of the components has to be able to work with each other.

Smart technologies will evolve more quickly if people didn't have to buy entire systems, because let's face it, most of us aren't as wealthy as Bill Gates. We also don't want to have 15 remote control devices for 15 different devices; we've been there and done that with televisions and recorders. What consumers want are add-on systems that are easy-to-use. What small manufacturers want is to be able to compete in this new marketplace.

"Two things are needed to make homes truly 'smart'," writes research journalist Ira Brodsky in *Computerworld*. "First are sensors, actuators, and appliances that obey commands and provide status information." These digital devices are already omnipresent in our appliances. "Second are protocols and tools that enable all of these devices, regardless of vendor, to communicate with each other," says Brodsky. This is the problem, but Brodsky believes that "smartphone apps, communication hubs, and cloud-based services are enabling practical solutions that can be implemented right now."

Home energy management systems (HEMS) have been the first wave of smart home devices, with hardware and software that monitors and controls a home's heating, ventilation, and air conditioning (HVAC) systems. As standards and protocols are being developed, the devices in our homes are making them appear smart.

Prototype Houses

The U.S. Department of Energy encourages new smart designs by sponsoring a Solar Decathlon, held every other year. Architecture and engineering college student teams compete in a number of categories, including intuitive control of devices and appliances. In 2013 a team from Canada described their engineering as an "integrated mechanical system" controlled by mobile devices. This is a student prototype of a smart home. Team Ontario's design for their house is called ECHO.

Domotics and Home Automation

As the smart house evolves, so, too, do the words we use to describe it. Most generally, home automation and home technology have been the early descriptors. Smart home automation has derived from those terms.

The word domotics literally means home robotics. In Latin, the word *domus* means home. The field of domotics encompasses all phases of smart home technology, including the highly sophisticated sensors and controls that monitor and automate temperature, lighting, security systems, and many other functions.

No need for those pesky robots, however. These days most mobile devices, like "smart" phones and tablets, are digitally connected and control many home systems. And what will your smart home look like? It should look just like what you're living in now if that's what you want.

(795 words)

https://www.thoughtco.com/what-is-a-smart-house-domotics-177572

New Words and Expressions

actuator [ˈæktjueitə] *n.*	执行器,执行元件;驱动器
android [ˈændrɔid] *n.*	人形机器人
antics [ˈæntiks] *n.*	滑稽可笑的举止;荒唐行为
domotics [dəuˈmɔtiks] *n.*	居家机器人;家庭自动化
havoc [ˈhævək] *n.*	大破坏,灾害
inventory [ˈinvəntɔːri] *v.*	开列清单
omnipresent [ˌɔmniˈpreznt] *a.*	无所不在的,遍及各处的
outlet *n.*	电源插座

Electrical Engineering PART 2

pesky [ˈpeski] *a.*	讨厌的,麻烦的
protocol [ˈprəutəkɔl] *n.*	(数据传递的)协议,规约
prototype [ˈprəutətaip] *n.*	原型
vendor [ˈvendə] *n.*	供应商
whirling [ˈwəːliŋ] *a.*	旋转的
communication hub	通讯集线器

Notes

1. Solar Decathlon [diˈkæθlən] 国际太阳能十项全能竞赛(由美国能源部发起和主办、以全球高校为参赛单位的太阳能建筑科技竞赛)

📖 Text Understanding

I. Fill in the blanks with proper words according to the passage.

1. The word domotics literally means _____.
2. _____ are sent through the home's wiring or sent wirelessly to switches and outlets that are programmed to operate appliances and electronic devices in every part of the house.
3. _____ have been the first wave of smart home devices.
4. Students majoring in _____ compete in Solar Decathlon every other year for new smart designs.

II. Give brief answers to the following questions.

1. Why is a smart home "intelligent"?
   _____
2. What's the name of the Gates' smart home?
   _____
3. What are the two things needed to make homes "smart"?
   _____
4. What was the prototype house in Team Ontario's design in Solar Decathlon 2013?
   _____

Vocabulary Building

III. Fill in the table below by giving the corresponding translation.

English	Chinese
	智能家居
	遥控设备
domotics	
communication hub	
	科幻小说
	原型
ventilation system	
	照明

Translation Practice

IV. Translate the following sentences from the passage into Chinese.

1. A smart house is a home that has highly advanced, automated systems to control and monitor any function of a house: lighting, temperature control, multi-media, security, window and door operations, air quality, or any other task of necessity or comfort performed by a home's resident.

2. A 1999 Disney movie titled *Smart House* presents the comical antics of an American family that wins a "house of the future" with an android maid who causes havoc.

3. The field of domotics encompasses all phases of smart home technology, including the highly sophisticated sensors and controls that monitor and automate temperature, lighting, security systems, and many other functions.

Electrical Engineering PART 2

In-Class Reading

Section B
Everything Explained about Electric Cars: Basics of an EV

With the pace at which EV (electric vehicle) technology is improving, we can say that within no time electric cars will become as common on roads as conventional gasoline vehicles. Since it's an emerging technology, many people are still skeptical about it, majorly, due to a lack of awareness. Also, we know that the masses are curious to know about electric cars. Considering that, we've tried to put together and explain everything in brief about electric vehicles.

What Is an EV or Electric Car?

An electric car is simply a four-wheel vehicle that runs on electric energy. In electric cars, electric energy is used to carry out mechanical operations. Unlike conventional vehicles, they don't require any internal combustion (IC) engine to operate. Hence, any kind of hydrocarbon fuel is not needed by electric cars. Due to the absence of an IC engine, it doesn't produce any harmful emissions. This is why electric vehicles are also known as green vehicles and considered better for the environment.

Components of Electric Cars

Since there is no engine, there are other components that altogether give life to an electric vehicle. They are listed as follows:

- Batteries / Battery Pack (Mostly Lithium-ion)
- Electric Motors
- Inverters
- Battery Management System (BMS)
- Electric Vehicle Transmission
- Charger for Electric Cars

1. What are electric car batteries?

Batteries are the powerhouse of electric vehicles. They store the electric energy required by an EV to spin its wheels. You can think of it as a fuel tank in conventional gas cars. Batteries are also one of the most expensive parts of electric cars. Most EV batteries (Fig. 5-2) are made up of lithium-ion because of their high energy density and other benefits. The size of the batteries plays an important role in determining the electric range and power output of an electric vehicle. Electric car batteries are rated in kWh.

Fig. 5-2 Electric vehicle batteries

2. What is an electric car motor?

Electric motors in EVs are responsible for generating power. They are also responsible for determining the total power output and the performance of an electric car. They are mounted in between the wheels of an EV. Based on the number of electric motors and their position (Fig. 5-3), an EV can be distinguished as either an FWD, RWD or an AWD. Electric motors are rated in kW.

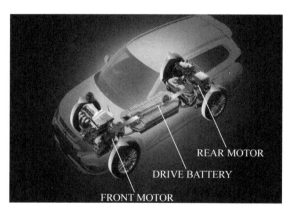

Fig. 5-3 The position of electric motors

3. Inverters in electric cars

Inverters in electric cars are used to convert the DC supply coming from the battery to an AC and then transfer it to the motor. After that, the motor generates the power and sends it to the wheels and car propels. Furthermore, in some EVs, when the brakes are applied, the kinetic energy (KE) is stored and used to charge the battery through a function called regenerative braking. Inverters play an important role in converting that KE into useful electric energy.

4. Battery management system (BMS) in electric vehicles

Battery Management System (BMS) in electric cars is an electronic system to look after the health of the batteries. BMS monitors various factors of an electric car battery like the voltage, temperature, current, etc. in order to prevent any damage and ensure a prolonged life.

5. Transmission system in electric cars / Do electric cars have gears?

Many people have doubts regarding the transmission system of electric cars. People often wonder if electric cars really have gears or not. Well, the answer to this question is both yes and no. Electric cars do have a transmission system, however, it's not like a normal gearbox you see in conventional vehicles. Electric cars generally have a single-speed transmission system. They don't have multiple gear ratios.

6. Electric vehicle charging socket

An electric car, like all electronic devices, needs a charger (Fig. 5-4) to charge itself. Electric cars have portable chargers that can be used to charge them at home or at DC fast charging stations.

Fig. 5-4 An electric vehicle charger

Working of an Electric Car

Electric vehicle technology is quite different from that of conventional cars. They have a different powertrain and components. Therefore, how these electric cars work is completely different from conventional cars.

However, it's not that complicated and you might even find it quite simple. As mentioned earlier, batteries in electric cars store energy. When you turn on the car, the inverter converts this energy into AC form and then transfers it to the motors. A single-speed transmission is used to control the amount being transferred to the wheels. This way the power is sent to the wheels and the electric car moves forward.

Do Electric Cars Use Oil?

Since there is no engine in electric vehicles, masses are skeptical whether EVs require any type of oil or not. Well, it's obvious that due to the absence of an engine, electric vehicles don't need engine oil.

However, there are other mechanical components that require periodic replacement of lubricants. For reference, brake fluid, transmission oil, thermal fluid, etc. are some of the common types of oil used in electric vehicles.

Advantages and Disadvantages of Electric Cars/Vehicles

Similar to most existing technologies, electric cars also have both advantages as well as some drawbacks. Let's outline some of the major pros and cons of electric vehicles in brief.

Benefits of electric cars

- Electric cars are better for the environment (zero emissions)
- Silent (very little or no noise pollution)
- Cheap and low maintenance
- Relatively spacious (more cargo space)
- Low running cost
- You can get tax credits and other incentives
- Quicker acceleration than gasoline cars

Drawbacks of electric cars

- Time consuming: charging electric vehicles requires significant time
- Most EVs have very short electric ranges
- Less available options for EVs in the market compared to gas vehicles
- Underdeveloped charging infrastructure in most countries
- Relatively expensive
- Battery chemicals can be harmful to the environment under certain conditions

Future of Electric Vehicles

Electric vehicles were first discovered back in the 18th century, but they never made it to the mainstream vehicles. EV technology has always been in cold storage until the present decade. The first modern electric car made its appearance in 1996 and it was introduced by GM. After that, it took almost 16 years for electric vehicles to explore the tip of their real potential. In 2012, Tesla made its debut in the industry and changed the whole scenario. After the entry of Tesla, there has been a revolution in the auto industry and nowadays there are multiple electric vehicles in the market. Several best electric cars are giving a tough time to their counterpart ICE vehicles.

With time, the numbers of EVs are increasing and people, as well as the government, are adopting them. Some prominent nations have already announced their deadlines for completely shifting towards electric vehicles, while other countries are also planning to shift towards electric mobility. One thing being clear, sooner or later, electric cars are going to dominate the automotive industry. What we need to see is how quickly it gets done and how EV technology will improve in the coming future.

(1176 words)

https://fossbytes.com/everything-explained-about-electric-cars/

Electrical Engineering PART 2

New Words and Expressions

charger *n.*	充电器
debut [deiˈbjuː] *n.*	初次登台;开张
gear [giə] *n.*	齿轮
gearbox *n.*	变速箱,齿轮箱
incentive [inˈsentiv] *n.*	激励,刺激
inverter [inˈvəːtə] *n.*	逆变器
lithium-ion [ˈliθiəm ˈaiən] *n.*	锂离子
lubricant [ˈluːbrikənt] *n.*	润滑油,润滑剂
powerhouse *n.*	动力源
scenario [səˈnɑːriəu] *n.*	设想,方案
socket [ˈsɔkit] *n.*	插座
battery pack	电池组
brake fluid	制动液,刹车油
fuel tank	燃油箱
gear ratio	齿轮比
pros and cons	正反两方面;有利有弊;赞成与反对
regenerative braking	再生制动
transmission system	传动系统;传输系统

Notes

1. AWD (All Wheel Drive)	全时四轮驱动,全时四驱
2. FWD (Front Wheel Drive)	前轮驱动,前驱
3. GM (General Motors)	通用汽车公司
4. ICE (internal combustion engine)	内燃机
5. kW (kilowatt)	千瓦,功率
6. kWh (kilowatt hour)	千瓦时,度电
7. RWD (Rear Wheel Drive)	后轮驱动,后驱

📖 Text Understanding

Ⅰ. Choose the best answer according to the passage.

1. In electric vehicles, electric energy is used to carry out _____ operations.

    A. electrical   B. mechanical   C. kinetic   D. kinematic
2. Electric vehicles are known as green vehicles, because _____.
    A. it produces zero emissions   B. it produces few emissions
    C. it produces some emissions   D. it produces many emissions
3. The following statements about electric vehicles are true except that _____.
    A. batteries are the powerhouse of electric vehicles
    B. electric motors are rated in kWh
    C. inverters play an important role in converting kinetic energy into useful electric energy
    D. electric vehicles are sold relatively expensive, but their maintenances are cheap
4. Electric vehicles have _____.
    A. a single-speed transmission system   B. a dual-clutch transmission system
    C. both A and B   D. either A or B
5. Electric vehicles use different types of oil except _____.
    A. brake fluid   B. transmission oil   C. thermal fluid   D. engine oil

II. Fill in the blanks with proper words according to the passage.

1. Unlike conventional vehicles, electric vehicles don't require any _____ engine to operate.
2. Most EV batteries are made up of lithium-ion because of their _____ and other benefits.
3. Based on the number of _____ and their _____, an EV can be distinguished as either an FWD, RWD or an AWD.
4. Battery Management System in electric cars is an electronic system to look after _____.
5. The first modern electric car made its appearance in _____ and it was introduced by _____.

III. Give brief answers to the following questions.

1. What are electric motors in EVs responsible for?
   _____
2. What is the function of inverters in electric cars?
   _____
3. What is the working principle of an electric car?
   _____
4. What are the attitudes of countries towards electric vehicles?
   _____

Electrical Engineering PART 2

📖 Vocabulary Building

IV. Fill in the table below by giving the corresponding translation.

English	Chinese
battery pack	
fuel tank	
	变速箱
	传动系统
RWD	
ICE	
	再生制动
	电动汽车

V. Match the items listed in the following two columns.

1. powertrain   a. a restraint used to slow or stop a vehicle

2. inverter   b. the group of components (engine, transmission, drive shafts, differentials, and the final drive) that generate power and deliver it to the road surface

3. brake   c. connected set of parts (clutch, gears, etc.) by which power is passed from the engine to the axle in a motor vehicle

4. charger   d. a substance capable of reducing friction by making surfaces smooth or slippery

5. lubricant   e. a device for charging or recharging batteries

6. transmission   f. a device to convert the DC supply from the battery to an AC

✍ Translation Practice

VI. Translate the following paragraph from the passage into Chinese.

Electric vehicles were first discovered back in the 18th century, but they never made it to the mainstream vehicles. EV technology has always been in cold storage until the present decade. The first modern electric car made its appearance in 1996 and it was introduced by GM. After that, it took almost 16 years for electric vehicles to explore the tip of their real potential. In 2012, Tesla made its debut in the industry and changed the whole scenario. After the entry of Tesla, there has been a revolution in the auto industry and nowadays there are multiple electric vehicles in the market. Several best electric cars are giving a tough time to their counterpart ICE vehicles.

Section C
How Does Quantum Levitation Work?

Quantum levitation as it is called is a process where scientists use the properties of quantum physics to levitate an object (specifically a superconductor) over a magnetic source (specifically a quantum levitation track designed for this purpose).

The Science of Quantum Levitation

The reason this works is something called the Meissner effect and magnetic flux pinning. The Meissner effect dictates that a superconductor in a magnetic field will always expel the magnetic field inside of it, and thus bend the magnetic field around it. The problem is a matter of equilibrium. If you just placed a superconductor on top of a magnet, then the superconductor would just float off the magnet, sort of like trying to balance two south magnetic poles of bar magnets against each other.

The quantum levitation process becomes far more intriguing through the process of flux pinning, or quantum locking, as described by Tel Aviv University superconductor group in this way:

Superconductivity and magnetic field do not like each other. When possible, the superconductor will expel all the magnetic field from inside. This is the Meissner effect. In our case, since the superconductor is extremely thin, the magnetic field DOES penetrate. However, it does that in discrete quantities called flux tubes. Inside each magnetic flux tube superconductivity is locally destroyed. The superconductor will try to keep the magnetic tubes pinned in weak areas (e.g. grain boundaries). Any spatial movement of the superconductor will cause the flux tubes to move. In order to prevent that, the superconductor remains "trapped" in midair.

The terms "quantum levitation" and "quantum locking" were coined for this process by Tel Aviv University physicist Guy Deutscher, one of the lead researchers in this field.

The Meissner Effect

A superconductor is a material in which electrons are able to flow very easily. Electrons flow through superconductors with no resistance, so that when magnetic fields get close to a superconducting material, the superconductor forms small currents on its surface, canceling out the incoming magnetic field. The result is that the magnetic field intensity inside of the surface of the superconductor is precisely zero. If you mapped the net magnetic field lines, it would show that they're bending around the object.

But how does this make it levitate? When a superconductor is placed on a magnetic track, the effect is that the superconductor remains above the track, essentially being pushed away by the strong magnetic field right at the track's surface. There is a limit to how far above the track it can be pushed, of course, since the power of the magnetic repulsion has to counteract the force of gravity.

A disk of a type-I superconductor will demonstrate the Meissner effect in its most extreme version, which is called "perfect diamagnetism", and will not contain any magnetic fields inside the material. It'll levitate, as it tries to avoid any contact with the magnetic field. The problem with this is that the levitation isn't stable. The levitating object won't normally stay in place.

In order to be useful, the levitation needs to be a bit more stable. That's where quantum locking comes into play.

Flux Tubes

One of the key elements of the quantum locking process is the existence of these flux tubes, called a "vortex". If a superconductor is very thin, or if the superconductor is a type-II superconductor, it costs the superconductor less energy to allow some of the magnetic field to penetrate the superconductor. That's why the flux vortices form, in regions where the magnetic field is able to, in effect, "slip through" the superconductor.

In the case described by the Tel Aviv team above, they were able to grow a special thin ceramic film over the surface of a wafer. When cooled, this ceramic material is a type-II superconductor. Because it's so thin, the diamagnetism exhibited isn't perfect ... allowing for the creation of these flux vortices passing through the material.

Flux vortices can also form in type-II superconductors, even if the superconductor material isn't quite so thin. The type-II superconductor can be designed to enhance this effect, called "enhanced flux pinning".

Quantum Locking

When the field penetrates into the superconductor in the form of a flux tube, it essentially turns off the superconductor in that narrow region. Picture each tube as a tiny non-superconductor region within the middle of the superconductor. If the superconductor moves, the flux vortices will move. Remember two things, though:
- the flux vortices are magnetic fields
- the superconductor will create currents to counter magnetic fields (i.e. the Meissner effect)

The very superconductor material itself will create a force to inhibit any sort of motion in relation to the magnetic field. If you tilt the superconductor, for example, you will "lock" or "trap" it into that position. It'll go around a whole track with the same tilt angle. This process of locking the superconductor in place by height and orientation reduces any undesirable wobble.

You're able to re-orient the superconductor within the magnetic field because your hand can apply far more force and energy than what the field is exerting.

Other Types of Quantum Levitation

The process of quantum levitation described above is based on magnetic repulsion, but there are other methods of quantum levitation that have been proposed, including some based on the Casimir effect. Again, this involves some curious manipulation of the electromagnetic properties of the material, so it remains to be seen how practical it is.

The Future of Quantum Levitation

Unfortunately, the current intensity of this effect is such that we won't have flying cars for quite some time. Also, it only works over a strong magnetic field, meaning that we'd need to build new magnetic track roads. However, there are already magnetic levitation trains (Fig. 5-5) in Asia which use this process, in addition to the more traditional electromagnetic levitation (maglev) trains.

Fig. 5-5 A magnetic levitation train

Another useful application is the creation of truly frictionless bearings. The bearing would be able to rotate, but it would be suspended without direct physical contact with the surrounding housing so that there wouldn't be any friction. There will certainly be some industrial applications for this, and we'll keep our eyes open for when they hit the news.

Quantum Levitation in Popular Culture

While the initial YouTube video got a lot of play on television, one of the earliest popular culture appearances of real quantum levitation was on the November 9 episode of Stephen Colbert's *The Colbert Report*, a satirical political pundit show on Comedy Central. Colbert brought scientist Dr. Matthew C. Sullivan from the Ithaca College Physics Department. Colbert explained to his audience the science behind quantum levitation in this way:

As I'm sure you know, quantum levitation refers to the phenomenon whereby the magnetic flux lines flowing through a type-II superconductor are pinned in place despite the electromagnetic forces acting upon them. I learned that from the inside of a Snapple cap.

He then proceeded to levitate a mini cup of his Stephen Colbert's Americone Dream ice cream flavor. He was able to do this because they had placed a superconductor disk within the bottom of the ice cream cup.

(1186 words)

https://www.thoughtco.com/quantum-levitation-and-how-does-it-work-2699356

New Words and Expressions

counteract [ˌkaʊntərˈækt] v.	抵消；中和；阻碍
diamagnetism n.	抗磁性；反磁性
equilibrium [ˌiːkwɪˈlɪbrɪəm] n.	平衡，均衡
film n.	薄膜
intriguing [ɪnˈtriːgɪŋ] a.	非常有趣的，引人入胜的
pundit [ˈpʌndɪt] n.	权威，专家，行家
quantum [ˈkwɒntəm] n.	量子
satirical [səˈtɪrɪkl] a.	讽刺的，讥讽的
vortex [ˈvɔːteks] n. pl. vortices	低涡，涡旋
wafer [ˈweɪfə] n.	晶圆
wobble [ˈwɒbl] n. & v.	摇晃
discrete quantities	离散量
electromagnetic levitation (maglev) train	磁悬浮列车
grain boundary	晶界
magnetic flux pinning	磁通钉扎
quantum levitation	量子悬浮
quantum locking	量子锁定

Notes

1. Snapple — 思乐宝(美国第一大果汁品牌)
2. Tel Aviv University — (以色列)特维拉夫大学
3. the Casimir [ˈkæsɪˌmɪr] effect — 卡西米尔效应

4. the Meissner ['maisnə] effect 迈斯纳效应

📖 Text Understanding

I. Decide whether the following statements are true (T) or false (F) according to the passage.

1. The Meissner effect makes the superconductor expel all the magnetic field from inside in any circumstance.
2. "Quantum levitation" and "quantum locking" were coined scientific terminology.
3. When a superconductor is placed on a magnetic track, the effect is that the superconductor remains above the track, but has a certain limit.
4. Type-I superconductor is so thin that some of the magnetic field is able to penetrate the superconductor.
5. If you tilt the superconductor, you will lock it into that position, because the superconductor has prohibited any sort of motion.
6. Both magnetic levitation cars and trains are available in the near future.

II. Give brief answers to the following questions.

1. What is quantum levitation?

2. What is the Meissner effect?

3. What is a superconductor?

4. Why is quantum locking necessary in application?

5. What is the working principle of frictionless bearings?

📖 Vocabulary Building

III. Fill in the table below by giving the corresponding translation.

English	Chinese
equilibrium	
discrete quantities	
	磁悬浮列车
	抗磁性
grain boundary	
counteract	
	涡旋
	量子悬浮

✎ Translation Practice

Ⅳ. Translate the following paragraph into Chinese.

Powerful electromagnets placed on both sides along the tracks of a railway produce magnetic buoyancy. With like magnetic poles facing each other all the time when the train moves, the train is levitated several inches above the rails, as a result of repellent magnetic fields between the train and the rails, which is known as "magnetic cushion". Since there is no direct contact between the wheels and the rails, and friction is zero, the train moves forward at a speed up to 300 miles per hour without any clang.

🗩 Translation Skill Ⅴ

词语的处理(2)—— 词性的转换

词性转换(conversion of parts of speech)是指在科技英语汉译过程中,为了使汉语译文符合英语原文的表述方式、方法和习惯,将英语原文中的某一词性转译成汉语的另一词性的翻译方法。词性转换在汉英翻译中较为常见,这是因为英汉两种语言对各类词语的使用习惯和频率均有所不同,例如,英语多用动词、名词,汉语多用动词。在翻译时,如果绝对地按照原文的词性进行翻译,会使得译文貌合神离、翻译腔十足。因此,在科技英语翻译过程中,只有对汉语译文进行适当调整和转换才能翻译出妥帖、通畅的译文。词性转换主要有以下几种情况。

1. 转换成汉语动词

英语中表达动作概念的动名词,具有动作意义的抽象行为名词,由动词派生的形容词,起形容词作用的现在分词和过去分词,具有动作意义的副词,以及小部分介词等均可根据具体情况转换成汉语动词。例如:

A. 名词转换成动词

Four wheel independently-actuated (4WIA) electric vehicle is a promising vehicle due to its potentials in emissions and fuel consumption reductions.

译文：四轮独立驱动电动车在降低排放和燃油消耗方面具有巨大的潜力，是一种很有发展前景的汽车。

B. 形容词转换成动词

Therefore, it is common practice to assume that deflections are negligible and parts are rigid when analyzing a machine's kinematic performance.

译文：因此，在分析机器的运动性能时，通常的做法是假定弯曲变形忽略不计，并且零件是刚性的。

C. 副词转换成动词

By selectively passing up to half the exhaust gas back through the engine, the amount of oxygen available for combustion can be reduced.

译文：通过选择把一半的尾气重新输入发动机，可以减少参与燃烧的氧气。

D. 介词转换成动词

A large segment of mankind turns to untrammelled nature as a last refuge from encroaching technology.

译文：很多人都将无拘无束的大自然视为躲避现代科技侵害的最后乐土。

2. 转换成汉语名词

在翻译时，虽然英语名词转换成汉语动词的概率更大一些，但英语动词有时也需要转换成汉语名词，特别是英语中一些由名词派生出来的动词（如 differ, behave, impress, symbolize 等），在汉语中往往找不到对应的动词，在这种情况下，英语动词需要转换成汉语名词。除此之外，英语中一些具有抽象性质的形容词、副词等在翻译时也可以转换成汉语名词。再者，英语中的代词在翻译时需将其指代对象具体化，因此也需要转换成汉语名词。例如：

A. 动词转换成名词

Gases differ from solids in that the former have greater compressibility than the latter.

译文：气体和固体的区别在于前者比后者有更大的可压缩性。

B. 形容词转换成名词

A bridge engineer must have three points in mind while working on a bridge project: 1) creative and aesthetics; 2) analytical; 3) technical and practical.

译文：桥梁工程师在设计桥梁时必须牢记三点：1)创新性和美观性；2)可分析性；3)较高的技术水平和实用性。

C. 副词转换成名词

Oxygen is one of the important elements in the physical world, very active chemically.

译文：氧是物质世界的重要元素之一,其化学性质很活泼。

D. 代词转换成名词

We need frequencies even higher than those we call very high frequency.

译文：我们所需要的频率,甚至比我们称作高频率的频率还要高。

3. 转换成汉语形容词、副词

众所周知,副词修饰动词、形容词修饰名词。在英译汉时,英语的动词被转换成了名词,那么修饰该动词的副词也相应地需要转换成形容词,反之亦然。例如：

A. 副词转换成形容词

Virtual technology is widely utilized in various vehicle test-beds.

译文：虚拟技术在各种汽车试验平台上得到广泛的应用。

B. 形容词转换成副词

Slight increase in the hydro carbon emission was observed.

译文：我们观察到碳氢化合物的排放略微地增加。

C. 其他词性转换成形容词、副词

This vehicle can travel up to 155mph, so audibility of the alert function at higher speeds is a necessity.

译文：该车车速可以达到每小时155英里,因此在更高速度下警报功能的可听性是必要的。

We shall develop the aircraft industry in a big way.

译文：我们将大规模地发展航空工业。

以上词性转换方法的介绍旨在打破英汉翻译的误区,即不能一味地认为在翻译时必须严格忠实于原文的词性,英语名词只能译成汉语名词,英语动词只能译成汉语动词等;但这一翻译方法的引介也并不是说一遇到上述情况就需要进行词性转换。在翻译时,我们需要在充分理解的基础上灵活变通,摆脱词性的羁绊,一切以译文能否自然流畅地传递意义为翻译标准。

PART III
Mechanical Engineering

Unit 6

Basic Mechanisms and Elements

Before-Class Reading

Answer the following question before reading the passage:

1. Have you any ideas of what a mechanism is in mechanical engineering? If you do, give some examples to clarify your understanding.

Section A
Brief Introduction to Mechanism

A mechanism is a device designed to transform input forces and movement into a desired set of output forces and movement. Mechanisms generally consist of moving components such as gears and gear trains, belt and chain drives, cam and follower mechanisms, and linkages as well as friction devices such as brakes and clutches, and structural components such as frames, fasteners, bearings, springs, lubricants, and seals, as well as a variety of specialized machine elements such as splines, pins, and keys.

A mechanism is an assembly of moving components. But sometimes an entire machine may be referred to as a mechanism. Examples are the steering mechanism in a car, or the winding mechanism of a wristwatch.

Multiple mechanisms are a machine. Machines have some or all of the four principal uses:

1. Transform energy A windmill transforms energy from the wind into mechanical energy to crush grain or electrical energy to power our homes.

2. Transfer energy The two gears in a can opener transfer energy from your hand to the edge of the can.

3. Multiply and/or change direction of force A system of pulleys can lift a heavy box up while you pull down with less effort than it would take to lift the box without help.

4. Multiply speed The gears on a bicycle allow the rider to trade extra force for increased speed, or sit back and pedal easily, at the expense of going slower.

(243 words)

From *Introduction to Mechanisms and Machines*

https://cdn.makezine.com/make/pdf/MTM_chap01.pdf

New Words and Expressions

assembly [əˈsembli] n.	装配,组装;集合
cam [kæm] n.	凸轮
clutch [klʌtʃ] n.	离合器
fastener [ˈfɑːsənə] n.	固定器,紧固零件
follower n.	从动件,从动轮
frame [freim] n.	框架;车架
linkage [ˈliŋkidʒ] n.	联动装置,连杆机构;连接
mechanism [ˈmekənizəm] n.	机制,机械装置
pedal [ˈpedl] v. & n.	踩踏板,骑车 & 踏板
pulley [ˈpuli] n.	滑轮,皮带轮
seal n.	密封垫
spline [splain] n.	花键
windmill n.	风车;风车磨坊
can opener	开罐器
chain drive	链条传动
gear train	齿轮传动链,轮系
machine element	机械零件,机械元件
steering mechanism	转向机制,转向装置
winding mechanism	绕线机制,发条

Exercises

📖 Text Understanding

Ⅰ. Give brief answers to the following questions.

1. What is a mechanism?

2. What do mechanisms consist of?

3. What are the principal applications of machines?

Vocabulary Building

II. Fill in the table below by giving the corresponding translation.

English	Chinese
	装配,组装
	润滑油
spline	
gear train	
	机械零件
	滑轮
winding mechanism	
clutch	
	连杆机构

In-Class Reading

Section B
Transmission

A transmission is a set of mechanism arranged to transfer rotational torque from one part of a mechanical system to another. Usually, mechanical transmitting systems are classified into gear trains, belt drives as well as chain drives according to their respective features.

Gear Trains

Gears are mechanical components that are designed to transmit rotational force to other gears at different speeds, torque, or in different directions. According to their construction and arrangement, gears of unequal diameters can be combined to produce a positive-ratio drive, so that the rotational speed and torque of the second gear are different from those of the first.

Gear trains are the most common type of all types of drives. They are used for transmitting rotary motion from one shaft to another with teeth. The accurately shaped teeth mesh with the teeth of another gear to provide positive-motion drive. There are many different types of gears. In the following part, we will sketch six basic types, namely spur gears, helical gears, bevel gears, worm gears, rack-and-pinion gears, and planetary gears.

Spur gears The most commonly used gear is called a spur gear. Spur gears are the simplest type. They are generally used on drives requiring moderate speed. They transmit motion between parallel shafts, as shown in Fig. 6-1.

Helical gears Helical gears have their tooth element at an angle or helix to the axis of the gear, as shown in Fig. 6-2. They are more difficult and expensive to make than spur gears, but are quieter and stronger. They may be used to transmit power between parallel shafts at an angle to each in the same or different planes.

Bevel gears Bevel gears are used to connect shafts, which are not parallel to each other. Usually the shafts are 90° to each other, but they may be more or less than 90°. A miter gear is a specific kind of bevel gear that is cut at 45° so that the two shafts end up at a 90° angle, as shown in Fig. 6-3.

Fig. 6-1 Spur gears   Fig. 6-2 Helical gears   Fig. 6-3 Bevel gears

The two meshing gears may have the same number of teeth for the purpose of changing direction of motion only, or they may have a different number of teeth for the purpose of changing both speed and direction.

Worm gears Worm gears actually look more like a screw than a gear, as shown in Fig. 6-4. They are designed to mesh with the teeth of a spur gear.

One important feature of the worm gear is the mechanical advantage it gives. When a worm gear (sometimes just called the worm) rotates one full revolution, the mating gear advances only one tooth. If the mating gear has 24 teeth, it gives the drive train a 24 : 1 mechanical advantage. Of course, the mating gear will be moving very slowly, but a lot of time. The trade-off is worth it.

Another great feature of worm gears is that in the majority of the time, they don't back drive. This means that the worm can turn the worm gear, but it won't work the other way around. The geometry and the friction just don't allow it. So, a worm gear drive train is desirable in positioning and lifting mechanisms where you don't want to worry about the mechanism slipping once a certain position is reached.

Rack-and-pinion gears A pinion is just another name for a spur gear, and a rack is a linear gear. A rack is basically a spur gear unwrapped so that the teeth lay flat, as shown in Fig. 6-5. The combination is used in many steering systems, and it is a great way to convert from rotary to linear motion. Movement is usually reciprocating, or back and forth, because the rack will end at some point, and the pinion can't push it in one direction forever.

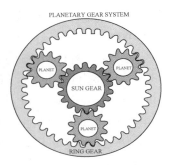

Fig. 6-4   Worm gears   Fig. 6-5   Rack and pinion   Fig. 6-6   Planetary gears

Another common example of a rack-and-pinion gear is a wine bottle opener. The rack in this case is circular, wrapped around the shaft that holds the corkscrew. The handles are a pair of first-class levers that end in pinion gears, and they go through a lot of travel when you push them down to give you the mechanical advantage needed to lift the cork out of the bottle easily.

Planetary gears Planetary or epicyclic gears, are a combination of spur gears with internal and external teeth, as shown in Fig. 6-6. They are mostly used in places where a significant mechanical advantage is needed, but there isn't much space, as in an electric screwdriver or an electric drill. You can even layer planetary gear sets to increase the mechanical advantage.

Gear ratios Gears of different sizes transmit a mechanical advantage, similar to how pulleys work. As always, the mechanical advantage is the ratio of how much we put into to how much we get out.

The smaller of two gears in a set is usually called a pinion, and is the one being driven. Let's say we have a 20-tooth pinion attached to a motor shaft. Then a 100-tooth spur gear mates with the pinion to rotate an adjacent shaft. The pinion must rotate five times to turn the output gear once, so the mechanical advantage is 5 : 1.

When the gear train is being used to magnify force, the input gear will always be smaller than the output gear. This setup is great when you have a motor and need to multiply the work it can do by itself, or when you need to slow the motor's output to a speed that fits your application.

To use a gear train to magnify speed, reverse the gears so the big gear is the input gear. The gear train is at a mechanical disadvantage in this configuration, but since one turn of the input gear on the motor turns the mating gear five times, the speed of the output is magnified by five.

So, take a look around you and see what kind of mechanisms you can find that have gears in them. How about the old clock, your blender, or a can opener? The kitchen is a great place to go looking for all sorts of useful mechanisms.

Belt Drives

Belt drive is a widely-used method of transmitting power from one shaft to another with pulleys and belts. They are not only used for parallel shafts but for crossed shafts as well. And they also feature simple construction, silent operation, less vibration and shocks in transmission. When compared with gear trains and chain drives, they are likely to slip when overloads are applied, and they occupy more space and larger center distance.

There are several typical belts for this kind of drive: open belt, cross belt, serpentine belt, and quarter-turn belts. Two popular types of belts used for the drives are flat belts (Fig. 6-7 (a)) and V-belts (Fig. 6-7 (b)).

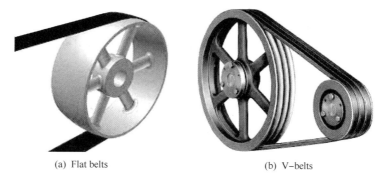

(a) Flat belts   (b) V-belts

Fig. 6-7　Different types of belts

Chain Drives

Similar to the open-belt drives, chain drives (Fig. 6-8) are fit for parallel shafts at larger center distances than those of gear drives. With a chain connecting sprockets (wheels) on the driving and driven shafts, both the input and output rotation of shafts can be driven in the same direction or in opposite direction. However, their velocity ratio transmitted from one shaft to another depends on the size of the two sprockets.

Fig. 6-8　Chain drives

Although chain drives are distinguished by good efficiency and no slip (unlike belt drives), the contact between the chain and the tooth on the sprockets is often noisy.

(1274 words)

From *Introduction to Mechanisms and Machines*

https://cdn.makezine.com/make/pdf/MTM_chap01.pdf

New Words and Expressions

configuration [kənˌfigəˈreiʃən] n.	配置,结构
corkscrew [ˈkɔːkskruː] n.	开瓶器,螺丝锥
drill n. & v.	钻子 & 钻孔
geometry [dʒiˈɔmitri] n.	几何学
helix [ˈhiːliks] n.	螺旋,螺旋状物
lever [ˈliːvə] n.	杠杆
linear [ˈliniə] a.	直线的,线性的
mesh v.	啮合
pinion [ˈpinjən] n.	小齿轮
rack n.	齿条
screw [skruː] n.	螺丝,螺钉,螺杆
screwdriver [ˈskruːdraivə] n.	螺丝刀
sprocket [ˈsprɔkit] n.	链轮齿,扣链齿轮
tooth n.	轮齿
velocity [vəˈlɔsəti] n.	速度,速率
worm n.	蜗杆
back drive	反向传动
bevel gear	伞齿轮,锥齿轮
cross belt	交叉皮带

epicyclic [ˌepiˈsaiklik] gear	行星齿轮
helical gear	斜齿轮
mechanical advantage	机械效益
miter gear	等径伞齿轮；斜方齿轮
open belt	开口皮带
planetary gear	行星齿轮机构
quarter-turn belt	直角回转皮带，直角挂轮皮带
rack-and-pinion gear	齿轮齿条式齿轮
serpentine [ˈsəːpəntain] belt	蛇纹岩带
spur gear	正齿轮，直齿轮
worm gear	蜗轮

📖 Text Understanding

Ⅰ. **Decide whether the following statements are true (T) or false (F) according to the passage.**

1. Mechanical transmission system can be classified into gear trains, belt drives, and chain drives.
2. Gears are used to transmit any motion from one shaft to another with teeth.
3. Spur gears are the simplest and the most commonly used gears.
4. Helical gears are quieter and stronger than spur gears.
5. Spur gears, helical gears, and bevel gears all have parallel shafts.
6. The meshing gears of bevel gears can either have the same or different number of teeth for different purposes.
7. A wine bottle opener is a good example of a rack-and-pinion gear.
8. If a 30-tooth pinion meshes with a 90-tooth spur gear, then the mechanical advantage should be 1∶3.
9. If we want to magnify force, the input gear should be the bigger gear.
10. Compared with gear trains and chain drives, belt drives are likely to slip when overloaded.

Ⅱ. **Fill in the blanks with proper words according to the passage.**

1. A transmission is a set of mechanism arranged to transfer _____ from one part of

a mechanical system to another.

2. Among gear trains, belt drives, and chain drives, _____ are the most common types of all.
3. A _____ is another name for a spur gear.
4. _____, a special kind of gears, serve the purpose of changing direction of motion only.
5. A worm gear drive train is desirable in _____ and _____ mechanisms.
6. The main components of a planetary gear are _____, _____, and _____.
7. When a 25-tooth pinion meshes with a 100-tooth gear, the mechanical advantage is _____.
8. Belt drives are used both for _____ shafts and _____ shafts.

III. Give brief answers to the following questions.

1. List at least five different types of gears. Which kind of gear is frequently used in steering mechanism?

2. Compared with spur gears, what are the merits and shortcomings of helical gears?

3. What are the two distinctive features of worm gears?

4. Where are planetary gears mostly used? Give some examples.

5. What are the advantages of chain drives?

IV. Identify different kinds of gears.

A   B   C   D

Vocabulary Building

V. Fill in the table below by giving the corresponding translation.

English	Chinese
	电钻
	螺丝刀
lever	
blender	
	线性运动
	往复运动
worm gear	
miter gear	
	速率比
	转矩,扭矩

VI. Match the items listed in the following two columns.

1. transmission
2. gear ratio
3. mechanical disadvantage
4. sprocket
5. pulley
6. rack

a. a state that the input gear is bigger than the output gear

b. a wheel on an axle or shaft that is designed to support movement and change of direction of a cable or belt along its circumstance

c. can be calculated directly from the numbers of teeth on the gears in the gear train

d. thin wheel with teeth that engage with a chain

e. basically a spur gear unwrapped so that the teeth lay flat

f. a set of mechanism arranged to transfer rotational torque from one part of a mechanical system to another

Translation Practice

VII. Translate the following sentences into Chinese.

1. Gears are mechanical components that are designed to transmit rotational force to other gears at different speeds, torque, or in different directions. According to their construction and arrangement, gears of unequal diameters can be combined to produce a positive-ratio drive, so that the rotational speed and torque of the second gear are different from those of the first.

2. Helical gears have certain advantages. For example, when connecting parallel shafts, they have a higher load-carrying capacity than spur gears with the same tooth numbers.

3. Similar to the open-belt drives, chain drives are fit for parallel shafts at larger center distances than those of gear drives. With a chain connecting sprockets on the driving and driven shafts, both the input and output rotation of shafts can be driven in the same direction or in opposite direction.

Section C
Basic Machine Elements

Machine element refers to an elementary component of a machine. These elements consist of three basic types:

(1) **Structural components** such as frame members, bearings, axles, splines, fasteners, seals, and lubricants;

(2) **Mechanisms** that control movement in various ways such as gear trains, belt or chain drives, linkages, cam and follower systems, including brakes and clutches;

(3) **Control components** such as buttons, switches, indicators, sensors, actuators, and computer controllers.

In the following part, a brief description of some basic machine elements will be introduced.

Bearings

A bearing can be defined as a member specifically designed to support moving machine components. The most common bearing application is the support of a rotating shaft that is transmitting power from one location to another. Since there is always relative motion between a bearing and its mating surface, friction is involved. In many instances, such as the design of pulleys, brakes, and clutches, friction is desirable. However, in the case of bearings, the reduction of friction is one of the prime considerations. Friction results in loss of power, the generation of heat, and increased wear of mating surfaces.

There are two basic types of bearings: sliding (plain/journal) bearings and rolling (antifriction) bearings.

Sliding bearings In sliding bearings, a film of lubricant separates the moving part from the stationary part. The amount of sliding friction in sliding bearings depends on the surface finishes, materials, sliding velocities, and type of lubricant used.

Fig. 6-9 Sliding bearings

Sliding bearings (Fig. 6-9) are the simplest bearings to construct, and considering the multitude of pin-jointed devices and structures in use, they are probably most commonly used. The major advantage of sliding bearings is the cost. They are much cheaper to produce.

Rolling bearings Rolling bearings (Fig. 6-10) operate with rolling elements (either balls or rollers), and hence rolling resistance, rather than sliding friction, predominates. The cause of rolling resistance is the deformation of mating surface of the rolling element and the raceway on which it rolls.

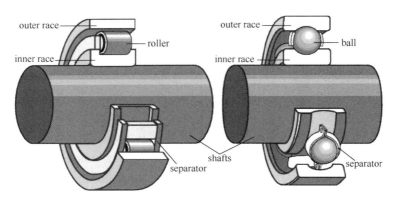

Fig. 6-10 Rolling bearings

A rolling bearing consists of two rings (inner ring and outer ring), the rolling bodies (balls, rollers or needles) and a separator. The roller runs in grooves of the rings. The bearings adjust themselves to small amount of angular misalignment in the assembled shaft and the separator keeps the rollers evenly spaced and prevents them from touching each other.

There are three types of rolling bearings: ball bearings (Fig. 6-11 (a)), roller bearings (Fig. 6-11 (b)), and needle bearings (Fig. 6-11 (c)). Ball and roller bearings can be designed to absorb thrust loads in addition to radial loads, whereas needle bearings are limited to radial load applications only.

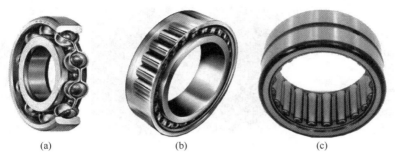

Fig. 6-11 Three types of rolling bearings

Springs

A spring is a load-sensitive and energy-storing device, the chief characteristics of which are an ability to tolerate large deformations without failure and to recover its initial size and shape when loads are removed. In general, a spring may be classified as either wire springs or flat springs, although there are variations within these divisions.

Wire springs Wire springs include helical (coil) springs (Fig. 6-12) of round, square, or special-section wire and are made to resist tensile, compressive or torsional loads.

Flat springs Flat springs include cantilever springs, leaf springs, Belleville washers (or Belleville springs), and the like. Cantilever springs are fixed only at one end and their application can be seen in Fig. 6-13. Leaf springs (Fig. 6-14 (a)) are used in vehicle suspension (Fig. 6-14 (b)), electrical switches, and bows. Belleville washers are characterized by short axial lengths and relatively small deformations, and are often used in stacks as illustrated in Fig. 6-15.

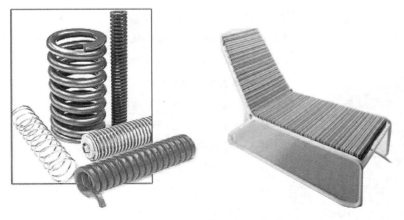

Fig. 6-12 Wire springs   Fig. 6-13 Application of cantilever springs

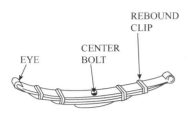  

(a) Leaf springs (b) Leaf springs in vehicle suspension

Fig. 6-14　Leaf springs and their application　　Fig. 6-15　Belleville washers

Although most springs are mechanical and derive their effectiveness from the flexibility inherent in metallic elements, hydraulic and air springs are obtainable.

Springs are used for a variety of purposes, such as supplying the motive power in clocks and watches, cushioning transport vehicles, measuring weights, restraining machine elements, mitigating the transmission of periodic disturbing forces from unbalanced rotating machines to the supporting structures, and providing shock protection for delicate instruments during shipments.

Couplings

A coupling is a mechanical device for uniting or connecting parts of a mechanical system. Couplings may be employed for a permanent or semi-permanent connection between shafts, or for disconnection of machine components to permit one member to run while the other is stationary.

Commercial shafts are limited in length by manufacturing and shipping requirements so that it is necessary to join sections of long transmission shafts with couplings. Couplings are also required to connect the shaft of a driving machine to a separately built driven unit. Permanent couplings are referred to simply as couplings, while those which may be readily engaged to transmit power, or disengaged when desired, usually are called clutches.

Fig. 6-16　A flange coupling

Rigid couplings　Rigid couplings are permanent couplings which by virtue of their construction have essentially no degree of angular, axial, or rotational flexibility. They must be used with collinear shafts.

The flange coupling shown in Fig. 6-16 is perhaps the most common coupling. It has the advantage of simplicity and low cost, but the connected shafts must be accurately aligned to avoid severe bending stress and excessive wear in the bearings. The length of the hub is determined by the length of the key required, and the hub diameter is approximately twice the bore. The thickness of the flange is determined by the permissible bearing pressure on the bolts. Although usually not critical, the shearing stress on the cylindrical area where the flange joints the hub should be checked.

Flexible couplings Flexible couplings are used to transmit torque from one shaft to another when the two shafts are slightly misaligned. Flexible couplings can accommodate varying degrees of misalignment up to 3° and some parallel misalignment. In addition, they can also be used for vibration damping or noise reduction. Fig. 6-17 shows the connection of flexible couplings.

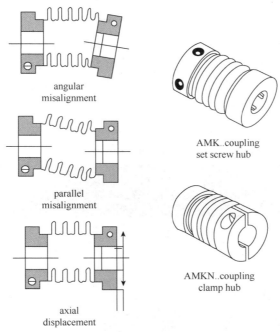

Fig. 6-17 The connection of flexible couplings

A beam coupling (Fig. 6-18), also known as a helical coupling, is a flexible coupling for transmitting torque between two shafts while allowing for angular misalignment, parallel offset, and even axial motion, of one shaft relative to the other. This design utilizes a single piece of material and becomes flexible by removal of material along a spiral path resulting in a curved flexible beam of helical shape. Since it is made from a single piece of material, the beam style coupling does not exhibit the backlash found in some multi-piece couplings. Another advantage of being an all machined coupling is the possibility to incorporate features into the final product while still keeps the single piece integrity.

Fig. 6-18 A beam coupling

Clutches

A clutch is a mechanical device that provides for the transmission of power from one component (the driving member) to another (the driven member) when engaged, but can be disengaged. In the simplest application, clutches are employed in devices which have two rotating shafts (drive shaft or line shaft). In these devices, one shaft is typically attached to a motor or other power unit (the driving member) while the other shaft (the driven member) provides output power for work to be done.

Clutches are most frequently used in vehicles. The following part will introduce clutches employed both as automobile powertrain and automobile non-powertrain.

Automobile powertrain In a modern car with a manual transmission, the clutch is operated by the left-most pedal using a hydraulic or cable connection from the pedal to the clutch mechanism. The default state of the clutch is engaged—that is the connection between engine and gearbox is always "on" unless the driver presses the pedal and disengages it. If the engine is running with clutch engaged and the transmission in neutral, the engine spins the input shaft of the transmission, but no power is transmitted to the wheels. The working principle of a dual-clutch transmission is illustrated in Fig. 6-19.

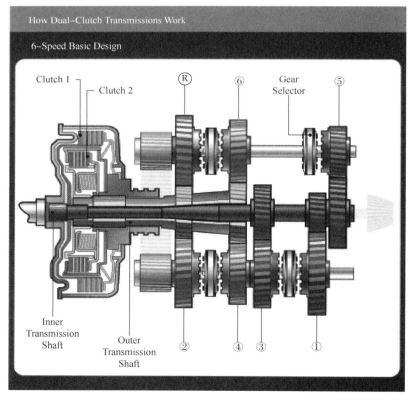

Fig. 6-19 Dual-clutch transmission

The clutch is located between the engine and the gearbox, as disengaging it is required to change gear. Although the gearbox does not stop rotating during a gear change, there is no torque transmitted through it, thus less friction between gears and their engagement dogs. The output shaft of the gear box is permanently connected to the final drive, then the wheels, and so both always rotate together, at a fixed speed ratio. With the clutch disengaged, the gearbox input shaft is free to change its speed as the internal ratio is changed. Any resulting difference in speed between the engine and gearbox is evened out as the clutch slips slightly during re-engagement.

Automobile non-powertrain There are other clutches found in a car. For example, a belt-driven engine cooling fan may have a clutch that is heat-activated. The driving and driven members are separated by a silicone-based fluid and a valve controlled by a bimetallic spring. When the temperature is low, the spring winds and closes the valve, which allows the fan to spin at about 20% to 30% of the shaft speed. As the temperature of the spring rises, it unwinds and opens the valve, allowing fluid past the valve which allows the fan to spin at about 60% to 90% of the shaft speed. Other clutches such as for an air conditioning compressor electronically-engaged clutches use magnetic force to couple the driving member to the driven member.

(1586 words)

From *Introduction to Mechanisms and Machines*

https://cdn.makezine.com/make/pdf/MTM_chap01.pdf

New Words and Expressions

backlash *n*.	后座,后冲
bimetallic [baimiˈtælik] *a*.	双金属的
bolt [bəult] *n*.	螺栓
bore *n*. & *v*.	镗孔,内径 & 钻孔,镗孔
bow [bəu] *n*.	弓;船头
coupling [ˈkʌpliŋ] *n*.	联轴器
cushion *v*.	起缓冲作用;(用垫子)使柔和
deformation [ˌdiːfɔːˈmeiʃən] *n*.	变形
diameter [daiˈæmitə] *n*.	直径
eye *n*.	吊环,孔
groove [gruːv] *n*.	槽
hub [hʌb] *n*.	中心,毂

misalignment [ˌmisəˈlainmənt] n.	未对准；移位
mitigate [ˈmitigeit] v.	使缓和，使减轻
pin-jointed a.	销接的
raceway n.	滚道
separator n.	轴承座
silicone [ˈsilikəun] n.	硅树脂，硅酮
suspension [səˈspenʃən] n.	汽车悬架，悬挂
valve [vælv] n.	阀门
angular misalignment	角度误差；角位移
antifriction bearing	减磨轴承，滚动轴承
axial [ˈæksiəl] length	轴向长度，轴长
ball bearing	滚珠轴承
beam coupling	电子束耦合，光束耦合
Belleville [ˈbelvil] washer/spring	蝶形弹簧，盘形弹簧
bending stress	弯曲应力
cantilever [ˈkæntiliːvə] spring	悬臂弹簧
change gear	换挡
coil spring	螺旋弹簧
collinear shaft	共线轴
compressive load	压缩载荷
default state	缺省状态
evenly spaced	均匀分布
flange coupling	凸缘联轴节
flat spring	平板弹簧，片弹簧
flexible coupling	弹性联轴器
helical spring	螺旋弹簧
journal bearing	滑动轴承；轴颈轴承
leaf spring	钢板弹簧（简称板簧）；弹簧片
line shaft	传动轴
manual transmission	手动变速箱
mating surface	啮合面，配合面
needle bearing	滚针轴承
plain bearing	滑动轴承；平面轴承
radial load	径向载荷
rebound clip	弹簧箍圈，弹簧连接夹
rigid coupling	刚性联轴器；固定耦合

roller bearing	滚柱轴承
rolling bearing	滚动轴承
shearing stress	剪应力
shock protection	防震保护
sliding bearing	滑动轴承
surface finish	表面抛光
tensile ['tensail] load	拉伸载荷
thrust load	轴向载荷,推力载荷
torsion load	扭力载荷
vibration damping	减振
wire spring	钢丝弹簧

Text Understanding

Ⅰ. Choose the best answer according to the passage.

1. In the following four choices, a(n) _____ is not the example of a control component.
   A. actuator   B. button   C. seal   D. sensor
2. Friction is desirable in the following circumstances except in the application of _____.
   A. brakes   B. bearings   C. clutches   D. pulleys
3. _____ bearings belong to a different type.
   A. Antifriction   B. Journal   C. Plain   D. Sliding
4. Needle bearings are designed to absorb _____ loads only.
   A. axial   B. radial   C. transverse   D. thrust
5. _____ springs are frequently used in vehicle suspension.
   A. Cantilever   B. Flat   C. Leaf   D. Wire
6. _____ can be readily engaged to transmit power, or disengaged when desired.
   A. Brakes   B. Clutches   C. Couplings   D. Suspensions
7. Flexible couplings can be used for _____.
   A. accommodating degrees of misalignment
   B. vibration damping

C. noise reduction

D. all of the above

II. Fill in the blanks with proper words according to the passage.

1. The three basic types of machine elements are _____, _____, and _____.

2. The amount of sliding friction in sliding bearings depends on _____, materials, _____, and type of lubricant used.

3. The three types of rolling bearings are _____, _____, and _____.

4. _____ are characterized by short axial lengths and relatively small deformations, and are often used in stacks.

5. A beam coupling is a flexible coupling for transmitting torque between two shafts while allowing for _____, _____, and even axial motion, of one shaft relative to the other.

6. The default state of the clutch is _____.

7. The clutch is located between _____ and _____.

8. An air conditioning compressor electronically-engaged clutches using _____ to couple the driving member to the driven member.

III. Give brief answers to the following questions.

1. What are the disadvantages of friction?
   _____

2. What are the chief characteristics of springs?
   _____

3. What is the most common coupling? What should be paid more attention to in its application?
   _____

4. Clutches are employed in devices which have two rotating shafts. What are the two shafts attached to respectively?
   _____

5. What is the working principle of a belt-driven engine cooling fan?
   _____

Vocabulary Building

IV. Fill in the table below by giving the corresponding translation.

English	Chinese
radial load	
surface finish	
	减振
	弯曲应力
powertrain	
separator	
	变形
	变速箱
cantilever spring	
default state	
	剪应力
	手动变速箱

Ⅴ. **Fill in the blanks with the words from the passage. The first letter of the word is given.**

1. A bearing can be defined as a member s_____ designed to support moving machine components.
2. The bearings adjust themselves to small amount of angular m_____ in the assembled shaft and the separator keeps the rollers evenly spaced.
3. Most springs are mechanical and derive their effectiveness from the f_____ inherent in metallic elements.
4. Powerful shock absorbers c_____ our landing.
5. The length of the hub is determined by the length of the key required, and the hub d_____ is approximately twice the bore.
6. You should d_____ the clutch before changing gear.
7. In a modern car with a manual transmission, the clutch is operated by the left-most pedal using a h_____ or cable connection from the pedal to the clutch mechanism.

✍ **Translation Practice**

Ⅵ. **Translate the following paragraphs into English.**

1. 轴承有各种类型,如滑动轴承、滚珠轴承、滚柱轴承等等。我们根据不同的需要,使用不同的轴承。尽管滚珠轴承和滚柱轴承的基本设计责任在轴承制造厂家,机器设计人员必须对轴承所要完成的任务做出正确的评价,不仅要考虑轴承的选择,而且还要考虑轴承的正确安装条件。

2. 传动装置、传动轴、轴承和其他部件都是典型的机械零件,广泛使用于各种不同的领域。传动装置用来降低或增加速度扭矩,以达到一个合适的输出功率。轴、轴承和其他部件也都用在各种机械和机械设备上。

Section D
Joints and Linkages

Mechanical system is to describe a system or a collection of rigid or flexible bodies that may be connected together by joints. The rigid bodies are called links. A linkage is a mechanical system consisting of links connected by either pin joints (also called revolute joints) or sliding joints (also called prismatic joints).

Links

A link is one of the rigid bodies or members joined together to form a kinematic chain. The term rigid link or sometimes simply link is an idealization used in the study of mechanisms that does not consider small deflection due to strains in the machine members. A perfectly rigid or inextensible link can exist only as a textbook type of model of a real machine member. For typical machine parts, maximum dimension changes are of the order of only a one-thousandth of the part length. We are justified in neglecting this small motion when considering the much greater motion characteristic of most mechanisms. The word link is used in a general sense to include cams, gears, and other machine members in addition to cranks, connecting rods, and other pin-connected components.

Joints

There are endless ways to join structural elements together to create movement, and each unique system of connections will give you a different type of motion. There are mainly four types of joints that are found in mechanisms:

(1) Revolute, rotary or pin joint (R)

(2) Prismatic or sliding joint (P)

(3) Spherical or ball joint (S)

(4) Helical or screw joint (H)

A revolute joint (Fig. 6-20) allows a rotation between the two connecting links. The best example of this is the hinge used to attach a door to the frame. A hinge will be able to move at most 180 degrees, most likely less depending on where your two structural pieces are hinged together. The rigidity of a hinge point is useful when you want to limit motion between two pieces to one plane.

A prismatic joint (Fig. 6-21) allows a pure translation between the two connecting links. Ideally, the sliding section of a prismatic joint should be at least twice as long as the width of the slot. This will minimize the chance of the piece binding up. The connection between a piston and a cylinder in an internal combustion engine or a compressor is via a prismatic joint.

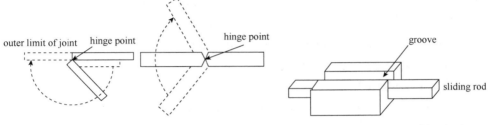

Fig. 6-20 A revolute (pin) joint   Fig. 6-21 A prismatic (sliding) joint

A spherical joint (Fig. 6-22) between two links allows the first link to rotate in all possible ways with respect to the second. The best example of this is seen in the human body. The shoulder and hip joints, called ball and socket joints, are spherical joints.

A helical joint allows a helical motion between the two connecting bodies. A good example of this is the relative motion between a bolt and a nut as shown in Fig. 6-23.

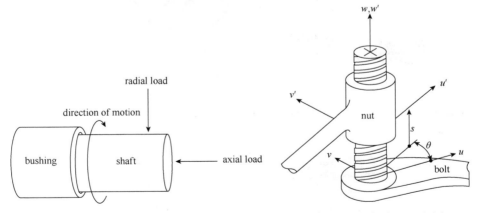

Fig. 6-22 A spherical (ball) joint   Fig. 6-23 A helical (screw) joint

Linkages

A linkage may be defined as an assemblage of solid bodies or links, in which each link is connected to at least two others by pin connections (hinges) or sliding joints (S-joints and H-joints are excluded). To satisfy this definition, a linkage must form an endless or closed chain or a series of closed chains. It is obvious that a chain with many links will behave differently from one with few. This raises the vitally important question regarding the suitability of a given linkage for the transmission of motion in a machine. This suitability depends on the number of links and the number of joints.

Perhaps the simplest linkage is the lever, which is a link that pivots around a fulcrum attached to ground, or a fixed point. Two levers connected by a rod so that a force applied to one is transmitted to the second are known as a four-bar linkage. The levers are called cranks, and the fulcrums are called pivots. The connecting rod is also called the coupler. The fourth bar in this assembly is the ground or frame, on which the cranks are mounted. A four-bar linkage is illustrated in Fig. 6-24 below.

The four-bar linkage is an adapted mechanical linkage used on bicycles (Fig. 6-25). With a normal full-suspension bike, the back wheel moves in a very tight arc shape. This means that more power is lost when going uphill. With a bike fitted with a four-bar linkage, the wheel moves in such a large arc that it is moving almost vertically. This way the power loss is reduced by up to 30%.

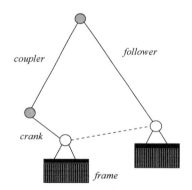

Fig. 6-24　A four-bar linkage with four R-joints

Fig. 6-25　A four-bar linkage used on bicycles

Linkages are important components of machines and tools. Examples range from the four-bar linkage used to amplify force in a bolt cutter or to provide independent suspension in an automobile, to complex linkage systems in robotic arms and walking machines. The internal combustion engine uses a slider-crank four-bar linkage (Fig. 6-26) formed from its piston, connecting rod, and crankshaft to transform power from expanding burning gases into rotary power. Relatively simple linkages are often used

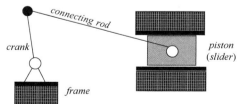

Fig. 6-26　A slide-crank linkage with three R-joints and one P-joint

to perform complicated tasks.

(865 words)

From *Analysis of Simple Planar Linkages*

by Kumar, V.

New Words and Expressions

assemblage [ə'semblidʒ] n.	装配,集合
bolt cutter	螺栓割刀;断线钳
bushing n.	轴衬,套管,衬套
crank [kræŋk] n.	曲柄
crankshaft n.	曲轴,机轴
deflection [di'flekʃn] n.	挠曲;偏差
fulcrum ['fuːlkrəm] n.	支点
hinge [hindʒ] n.	铰链
joint n.	关节,接合点
kinematic [ˌkinə'mætik] a.	运动的;运动学的
link n.	连杆
nut [nʌt] n.	螺母,螺帽
piston ['pistən] n.	活塞
slider-crank n.	曲柄滑块
strain [strein] n.	应变
translation n.	平移
ball and socket joint	球窝关节
connecting rod	连杆
flexible body	柔性体,弹性体
prismatic [priz'mætik] joint	移动关节
revolute joint	转动关节
rigid body	刚性体,刚体
screw joint	螺纹接头,螺旋接头
sliding joint	滑动关节
spherical ['sferikl] joint	球面关节

Text Understanding

I. Decide whether the following statements are true (T) or false (F) according to the passage.

1. There is no perfectly rigid or inextensible link in practical application.
2. A hinge in a revolute joint is able to rotate at any degree.
3. The sliding section of a prismatic joint must be at least twice as long as the width of the slot.
4. A helical joint between two links allows the first link to rotate in all possible ways with respect to the second.
5. A linkage is an assemblage of one or several of the four joints.
6. The lever is a link that pivots around a fulcrum attached to ground or a fixed point.

II. Give brief answers to the following questions.

1. What's the relationship between joints, links, and a linkage?

2. What are the four types of joints and their respective examples?

Vocabulary Building

III. Match the items listed in the following two columns.

1. rigid body	a. L-shaped bar and handle for converting to-and-fro movement to circular movement
2. deflection	b. the property of being bent
3. strain	c. a system of components assembled together for a particular purpose
4. hinge	d. an idealization of a solid body in which deformation is neglected
5. assemblage	e. a joint that holds two parts together so that one can swing relative to the other
6. crank	f. deformation of a physical body under the action of applied forces

Translation Practice

IV. Translate the following sentences from the passage into Chinese.

1. A link is one of the rigid bodies or members joined together to form a kinematic chain. The term rigid link or sometimes simply link is an idealization used in the study of mechanisms that does not consider small deflection due to strains in the machine members.
_____
_____

2. A linkage may be defined as an assemblage of solid bodies or links, in which each link is connected to at least two others by pin connections or sliding joints. To satisfy this definition, a linkage must form an endless or closed chain or a series of closed chains.
_____
_____

3. The four-bar linkage is an adapted mechanical linkage used on bicycles. With a normal full-suspension bike the back wheel moves in a very tight arc shape. This means that more power is lost when going uphill. With a bike fitted with a four-bar linkage, the wheel moves in such a large arc that it is moving almost vertically. This way the power loss is reduced by up to 30%.
_____
_____
_____

Translation Skill VI

词语的处理(3)—— 增词与省略

翻译时,基于忠实原则,译者不应对原文的内容随意增减。但由于英汉两种语言文字之间存在巨大的差异,在实际翻译过程中很难做到词句上完全对应。因此,为了准确传达出原文的信息,往往需要对译文作一些增减,但这里的增减绝不是无中生有或随意删除,而是确有需要才进行必要的增减。下面,我们将从增译法和省译法两方面分别进行举例说明。

1. 增译法

增译法(amplification)是指译者在英译汉时,在原文基础上添加必要的单词、词组、分句等，从而使译文在语气上、语义上、语法结构上、修辞表达上与原文保持一致，以达到译文与原文在内容、形式和思想方面的对等。

A. 增加表示复数概念的词语

<u>Signs</u> that repair is needed are transverse joint faulting, corner breaks, and lines at or near joints.

译文：需要修补的<u>各种现象</u>包括横向接头损坏、角部断裂和接头处或附近出现的裂缝。

B. 增加表示时态的词语

Worldwide, researchers <u>are working</u> to develop more efficient drive systems for vehicles, in particular using electric motors.

译文：世界范围内，许多研究者<u>正致力于</u>开发更高效的车用驱动系统，特别是使用电机的驱动系统。

C. 增加表示主语的词语

Based on a first order method, the sensitivity of the failure probability with respect to the random input quantities is evaluated.

译文：基于一阶矩法，<u>本文</u>/本研究对随机输入量的失效概率敏感性进行评价。

D. 增加具体和明确化的词语

But because <u>the pump</u> is driven by the engine, the injection pressure rises and falls along with the engine speed.

译文：但因为<u>输油泵</u>由发动机驱动，所以喷油压力随着发动机的转速起伏。

E. 增加概括性词语

The factors, voltage, current, and resistance are related to each other.

译文：电压、电流和电阻这<u>三种因素</u>是相互关联的。

2. 省译法

省译法（omission）是指原文中有些词在译文中略去不译，因为译文中虽无其词但已有其意，或者在译文中是不言而喻的。换言之，省译法是删去一些可有可无的，或者有了反而累赘或违背译文习惯表达的词，但并不是把原文的某些思想内容删去。

A. 冠词的省略

<u>The</u> aim of this system is to warn the vehicle's driver and reduce the reaction time in case an emergency brake is necessary.

译文：本系统的目的是在紧急制动情况下警告汽车驾驶者，减少驾驶者的反应时间。

B. 代词的省略

Friction always manifests <u>itself</u> as a force that opposes motion.

译文：摩擦总是表现为一种对抗运动的力。

C. 连词的省略

<u>As</u> the temperature increases, the volume of water becomes greater.

译文:温度增高,水的体积就增大。

D. 动词的省略

Second, the monorail makes much less noise than a rail vehicle because its wheels consist of rubber tires.

译文:第二点,由于单轨车辆采用橡胶轮胎,所以它的噪声比铁路车辆小。

E. 名词的省略

It is essential that the mechanic or technician understand well the characteristics of electric circuits.

译文:很好地了解电路特性对技术人员来讲十分重要。

综上所述,增译可以增加表示复数的词、表示时态的词、主语、具体化的名词、概括词等;而省译则经常省略冠词、代词、连词,有时还会省略不具有实际意义的动词或重复性的名词等。在翻译时需根据英汉两种语言在词汇、语法和修辞上的差异,灵活使用增减翻译技巧,尽可能地体现科技译文的精准性、客观性、衔接性和逻辑性。

Unit 7

Engineering Materials

Before-Class Reading

Answer the following question before reading the passage:
1. Can you name some physical properties as well as some mechanical properties of engineering materials? If you do, please list them out in detail.

Section A
General Properties of Engineering Materials

The principal properties of materials which are of importance to engineers are selecting materials. These properties can be broadly divided into two categories, namely the physical properties and the mechanical properties.

Physical Properties of Materials

Physical properties are concerned with such properties as melting temperature, electrical conductivity, thermal conductivity, density, corrosion resistance, magnetic properties, etc. The most important properties will be considered as follows.

Density Density is defined as mass per unit volume for a material. The derived unit usually used by engineers is kg/m^3. Relative density is the density of a material compared with the density of water at 4℃.

Electrical conductivity Electrical conductivity measures a material's ability to conduct an electric current. The electrical conductivity of a material varies dramatically in temperature. For example, metallic conductors of electricity all increase in resistance as their temperatures rise. Pure metals show this effect more strongly than alloys. However, pure metals generally have a better conductivity than alloys at room temperature.

Melting temperatures Melting temperatures and recrystallization temperatures have a great effect on the application of metals and alloys.

Semiconductors The conductivity of semiconductor materials increases rapidly for relatively slight temperature increase. Semiconductor materials are capable of having their conductors' properties changed during manufacture. Examples of semiconductor materials are silicon and germanium. They are used extensively in the electronics industry for the manufacture of solid-state devices such as diodes, transistors, and integrated circuits.

Thermal conductivity Thermal conductivity is the ability of a material to transmit heat energy by conduction. Fig. 7-1 shows a soldering iron. The bit is made from copper which is a good conductor of heat and it allows the heat energy stored in it to travel easily down to the tip and into the work being soldered. The wooden handle remains cool as it has a low thermal conductivity and resists the flow of heat energy.

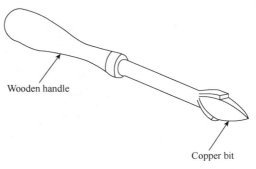

Fig. 7-1 Thermal conductivity

Fusibility Fusiblity is the ease with which materials will melt. It can be seen from Fig. 7-2 that solders melt easily, and therefore have the property of high fusibility. On the other hand, fire bricks used for the furnace linings only melt at very high temperatures, and therefore have the properties of low fusibility. Such materials which only melt at very high temperatures are called refractory materials.

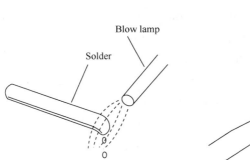
A solder melts at a low temperature, and therefore has a high fusibility.

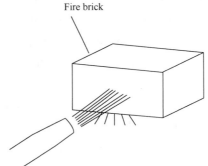
A fire brick melts at a very high temperature, and therefore has a low fusibility, and is called a refractory material.

Fig. 7-2 Fusibility

Magnetic reluctance Just as some materials are good or bad conductors of electricity, some materials can be good or bad conductors of magnetism. The resistance to magnetic circuit is referred to as magnetic reluctance. Good magnetic conductors have low reluctance and examples are ferromagnetic materials which get their name from the fact that they are made from iron, steel, and associated alloying elements, such as cobalt and nickel. All other materials are non-magnetic and offer high reluctance to the magnetic flux field.

Temperature stability Any changes in temperatures can have very significant effects on the structures and properties of materials. Creep is one of the typical examples.

Creep is defined as the gradual extension of a material over a long period of time whilst the applied load is kept constant. It is also an important factor for plastic materials, and it must be considered harmful to metals when they work continuously at high temperatures; for example, gas-turbine blades. The creep rate increases when the temperature rises, but becomes slower when the temperature drops.

Mechanical Properties of Materials

Mechanical properties are measures of how materials behave under applied loads. Another way of saying this is how strong a metal is when it comes in contact with one or more forces. The main properties involve the following: tensile strength, toughness, malleability, hardness, ductility, stiffness, brittleness, elasticity, and plasticity.

Tensile strength (TS) Strength is the ability of a material to resist applied forces without fracturing. Strength properties are commonly referred to as tensile strength, compressive strength, torsional strength, shear strength, fatigue strength, and impact strength.

Tensile strength is the ability of a material to withstand tensile (stretching) loads without breaking. For example, Fig. 7-3 shows a heavy load being held up by a rod fastened to beam. As the force of gravity acting on the load is trying to stretch the rod, the rod is said to be in tension. Therefore, the material from which the rod is made needs to have sufficient tensile strength to resist the pull of the load.

Toughness Toughness is the ability of a material to withstand bending or it is the application of shear stresses without facture. Rubbers and most plastic materials do not shatter, therefore they are tough. For example, if a rod is made of high-carbon steel, then it will bend without breaking under the impact of a hammer, while if a rod is made of cast iron, then it will be broken by impact load as shown in Fig. 7-4.

Malleability Malleability is the capacity of a material to withstand deformation under compression without rupture. A malleable material allows a useful amount of plastic deformation to occur under compressive load before fracture occurs. It is required for manipulation by such processes as forging, rolling, and rivet heading.

Hardness Hardness is the ability of a material to withstand scratching (abrasion) or indentation by another hard body. It is an indication of the wear resistance of a material. For example, Fig. 7-5 shows a hardened steel ball being pressed first into a hard material and then into a soft material by the same load. It can be seen that the ball only makes a small indentation in the hard material, but it makes a very much deeper impression in the soft material.

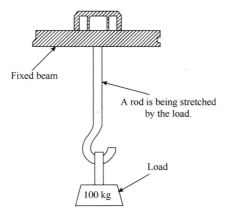

Fig. 7-3 Tensile strength

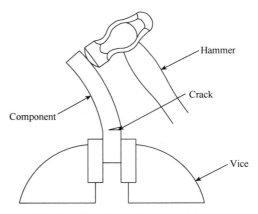

Fig. 7-4 Toughness (impact resistance)

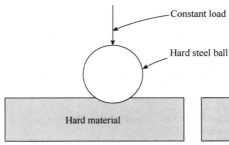

Fig. 7-5 Hardness

Ductility Ductility refers to the capacity of a material to undergo deformation under tension without rupture as in wire drawing or tube drawing (Fig. 7-6) operation.

Stiffness Stiffness is the measure of a material's ability not to deflect under an applied load. For example, although steel is much stronger than cast iron, cast iron is preferred for machine beds and frames because it is more rigid and less likely to deflect, which may bring consequent loss of alignment and accuracy.

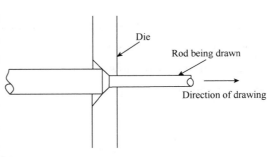

Fig. 7-6 Ductility

Consider Fig. 7-7 (a): For a given load, the cast iron beam deflects less than the steel beam because cast iron is a more rigid material. However, when the load increases as shown in Fig. 7-7 (b), the cast iron beam will break, whilst the steel beam deflects a little further but does not break. Thus a material which is rigid is not necessarily strong.

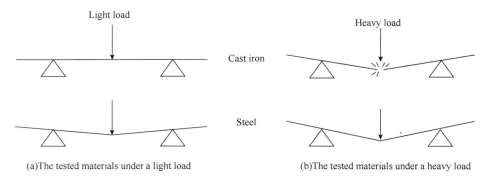

Fig. 7-7 Stiffness (rigidity)

Brittleness　Brittleness is the property of a material that shows little or no plastic deformation before fracture when a force is applied. Also, it is usually said as the opposite of ductility and malleability.

Elasticity　Elasticity is the ability of a material to deform under load and return to its original size and shape when the load is removed. An elastic material will be the same length when the load is removed, despite the fact that it will be longer whilst the load is being applied. All materials possess elasticity to some degree and each has its own elastic limit.

Plasticity　Plasticity is the exact opposite of elasticity, while ductility and malleability are particular cases of plasticity. It is the state of a material which has been loaded beyond its elastic limit so as to cause the material to deform permanently. Under such conditions, the material takes permanent set and will not return to its original size and shape when the load is removed. When a piece of mild steel is bent at right angles into the shape of a bracket, it shows the property of plasticity since it does not spring back strength again, as shown in Fig. 7-8.

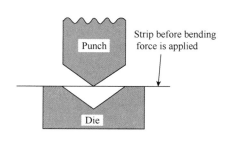
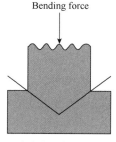

Fig. 7-8 Plasticity

Some metals such as lead have a good plastic range at room temperature and can be extensively worked (where working of metal means squeezing, stretching or beating it to shape). This is an advantage for plumbers when beating lead flashings to shape on building sites.

(1357 words)

https://www.electrical4u.com/mechanical-properties-of-engineering-materials

New Words and Expressions

abrasion [əˈbreiʒn] n.	(表层)磨损
bracket [ˈbrækit] n.	支架,托架
brittleness n.	脆性,脆度
cobalt [ˈkəubɔːlt] n.	钴
creep n.	蠕变(缓慢变形)
die n.	模具,冲模,钢模
ductility [dʌkˈtiliti] n.	延展性;柔软性;塑性
elasticity [ˌiːlæˈstisəti] n.	弹性,弹力
ferromagnetic [ˌferəumægˈnetik] a.	铁磁的
flashing n.	防水板,遮雨板
forging n.	锻造;锻件
fracture [ˈfræktʃə] v.	折断,断裂
fusibility [ˌfjusiˈbiliti] n.	可熔性,熔度
germanium [dʒəːˈmeiniəm] n.	锗
indentation [ˌindenˈteiʃən] n.	压痕,刻痕
malleability [ˌmæliəˈbiləti] n.	展延性;可锻性
malleable [ˈmæliəbəl] a.	有展延性的;可锻的
nickel [ˈnikl] n.	镍
plasticity [plæˈstisəti] n.	塑性,可塑性
plumber [ˈplʌmə] n.	管道工
punch [ˈpʌntʃ] n.	冲头,冲压机
recrystallization [riːkristəlaiˈzeiʃn] n.	重结晶,再结晶
refractory [riˈfræktəri] a. & n.	难熔的 & 耐火物质

rolling *n*.	轧制
rupture [rʌptʃə] *v*.	破裂,断裂
scratch [skrætʃ] *v*.	划损,刮坏
shatter [ˈʃætə] *v*.	破碎,粉碎
solder [səuldə] *n*. & *v*.	焊料 & 焊接
stiffness *n*.	刚度,挺度
toughness *n*.	韧性,韧度
vice [vais] *n*.	老虎钳
applied load	外加载荷,外施载荷
cast iron	铸铁;生铁
corrosion resistance	耐腐蚀性,抗腐蚀性
derived unit	导出单位
fatigue [fəˈtiːg] strength	疲劳强度
fire brick	耐火砖,防火砖
furnace lining	炉衬
impact strength	冲击强度
plastic deformation	塑性变形
rivet heading	铆钉打头
soldering iron	烙铁,焊铁
tube drawing	管材拉拔,拔管
wear resistance	耐磨性,耐磨度
wire drawing	拉线,拉丝

📖 Text Understanding

I. Choose the best answer according to the passage.

1. Among the following four general properties, _____ is not the physical property of engineering materials.

   A. density   B. ductility

   C. conductivity   D. fusibility

2. At room temperature, alloys usually have _____ pure metals.

A. the same conductivity as   B. a better conductivity than
C. a weaker conductivity than   D. no relations in conductivity with

3. A soldering iron mainly exhibits the physical property of _____.
   A. thermal conductivity   B. electrical conductivity
   C. density   D. temperature stability

4. A high-carbon rod steel will bend rather than break under the impact of a hammer, so it has _____ toughness.
   A. no   B. little   C. weak   D. strong

5. Cast iron is preferred for _____ because it is more rigid and less likely to deflect.
   A. machine beds   B. frames
   C. both A and B   D. neither A nor B

6. _____ is usually said as the opposite of ductility and malleability.
   A. Toughness   B. Hardness
   C. Brittleness   D. Plasticity

II. Match the term in the left column with the definition in the right column.

1. tensile strength   a. It is the ability of a material to withstand bending or it is the application of shear stresses without facture.

2. hardness   b. It is the measure of a material's ability not to deflect under an applied load.

3. ductility   c. It is the ability of a material to withstand tensile loads without breaking.

4. stiffness   d. It is the ability of a material to withstand scratching or indentation by another hard body.

5. elasticity   e. It is the ability of a material to deform under load and return to its original size and shape when the load is removed.

6. toughness   f. It is the capacity of a material to undergo deformation under tension without rupture.

III. Give brief answers to the following questions.

1. What is the relative density of a material?

2. What is fusibility? Please give an example of a material with a low fusibility.

3. What is creep? What's the relationship between the creep rate and the temperature change?

4. What is the distinct difference between malleability and ductility?

5. Taking elastic range and plastic range into consideration, what will happen to a material when the applied load is removed?

Vocabulary Building

IV. Fill in the table below by giving the corresponding translation.

English	Chinese
malleability	
rolling	
	刚度
	铸铁
corrosion resistance	
flashing	
	疲劳强度
furnace lining	
tube drawing	

Translation Practice

V. Translate the following sentences from the passage into Chinese.

1. Good magnetic conductors have low reluctance and examples are ferromagnetic materials which get their name from the fact that they are made from iron, steel, and associated alloying elements, such as cobalt and nickel.

2. Creep is defined as the gradual extension of a material over a long period of time whilst the applied load is kept constant.

_____
_____

3. Malleability is the capacity of a material to withstand deformation under compression without rupture. A malleable material allows a useful amount of plastic deformation to occur under compressive load before fracture occurs.

_____
_____
_____

4. Some metals such as lead have a good plastic range at room temperature and can be extensively worked. This is an advantage for plumbers when beating lead flashings to shape on building sites.

_____
_____
_____

In-Class Reading

Section B
Metallic Materials

Almost every substance known to man has found its way into the engineering workshop at some time or other. The most convenient way to study the properties and uses of engineering materials is to classify them into "families" as shown in Fig. 7-9 below.

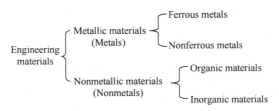

Fig. 7-9 Classification of engineering materials

Metals are divided into two general types—ferrous and nonferrous. Ferrous metals are those which contain iron and they can further be classified into "families" as shown in Fig. 7-10. Nonferrous metals are those which do not contain iron. However, some nonferrous metals may contain a small amount of iron as an impurity.

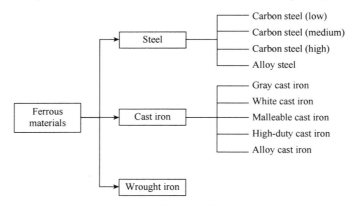

Fig. 7-10 Classification of ferrous metals

Steel and cast iron are the most common ferrous metals in general use. Steel is an alloy containing chiefly iron, carbon, and certain other elements in varying amounts. A wide range of physical properties may be obtained in steel by controlling the amount of carbon and other alloying elements and by subjecting the steel to various heat treatments.

Plain carbon steels usually contain, besides iron and carbon, a small amount of silicon, sulphur, phosphorus, and manganese. Alloy steels are formed by the addition of one or more of the following elements: nickel, chromium, molybdenum, vanadium, tungsten, manganese, silicon, and a small amount of other alloying elements.

Carbon is by far the most important alloying element in steel. It is the amount of carbon present that largely determines the maximum hardness obtainable. The higher the carbon content, the higher the tensile strength and the greater the hardness to which the steel may be heat-treated.

Low-carbon steels are usually used for low-strength parts requiring a great deal of forming. Medium-carbon steels are used for forgings and other applications where increased strength and a certain amount of ductility are necessary. High-carbon steels are used for high-strength parts such as springs, tools, and dies. Table 7-1 is a classification of ferrous materials according to their carbon content.

Carbon content of ferrous materials   Table 7-1

Ferrous materials	Carbon content
Wrought iron	Trace to 0.08%
Low-carbon steel	0.08% to 0.30%
Medium-carbon steel	0.30% to 0.60%
High-carbon steel	0.60% to 1.70%
Cast iron	1.70% to 4.5%

Alloy steels have special properties determined by the mixture and the amount of other metals added. To the metallurgist who works in metal mining and manufacturing, steels containing very small quantities of elements other than carbon, phosphorus, sulphur, and silicon are known as alloy steels. Each alloy steel has its own characteristic. A car is approximately made of 100 different kinds of alloy steels.

Some of the common alloying elements are described below:

Manganese Manganese helps to reduce certain undesirable effects of sulphur by combining with the sulphur. It also combines with carbon to increase hardness and toughness. Manganese possesses the property of aiding in increasing the depth of hardness penetration. It also improves the forging qualities by reducing brittleness at rolling and forging temperatures.

Silicon Silicon does not normally occur in steels in excess of 3.00 percent. A small amount of silicon improves ductility. It is used largely to increase impact resistance when combined with other alloys.

Sulphur Sulphur is generally regarded as detrimental to the hot working of steel and to the impact properties of steel treated to high tensile strength. However, sulphur is an invaluable aid to machining, and steels are often resulphurized to as high as 0.30 percent to gain advantage of this property.

Phosphorus Phosphorus has an undesirable effect on steel in that it imparts brittleness. There is some evidence that a small amount of phosphorus, less than 0.05 percent, increases tensile strength.

Nickel Nickel dissolves easily in molten steel. It is present in common nickel steels in a proportion of 0.40 percent up to 5.00 percent. The addition of nickel increases strength, yield point, hardness, and ductility. It also increases the depth of hardening. Nickel steels are less susceptible to warping and scaling than most other steels. Nickel increases corrosion resistance and is one of the major constituents of the "stainless" or corrosion-resisting steels.

Chromium Addition of chromium imparts hardness, strength, wear resistance, heat resistance, and corrosion resistance to steels.

Molybdenum Molybdenum, even in extremely small amounts, has considerable effect as an alloying element on the physical properties of steels. Molybdenum increases elastic limit, impact strength, wear resistance, and fatigue strength. Molybdenum steels are readily heat-treated, forged, and machined.

Vanadium Vanadium is usually used in amounts of less than 0.25 percent. As an alloying agent, vanadium improves fatigue strength, ultimate strength, yield point, toughness, and resistance to impact and vibration. Chromium-vanadium steels have good ductility and high strength.

Tungsten   Tungsten is used largely with chromium as a high-speed tool steel which contains 14.00 to 18.00 percent tungsten and 2.00 to 4.00 percent chromium. This steel possesses the characteristic of being able to retain a sharp cutting-edge even though heated to redness in cutting.

Tool-and-die steels   Tool-and-die steels are a large group of steels used when careful heat-treating must be done. These steels are used for parts such as chisels, hammers, screwdrivers, springs, and tools and dies to cut and form metals.

Tool steels with certain alloying elements are designed for specific uses. The most common kinds of tool steels include high-speed tool steels, hot work tool steels, cold work tool steels, and special-purpose tool steels.

Rolled steels   Rolled steels, which include bar, rod, and structural steels, are produced by rolling the steels into shape. Hot-rolled steels are formed into shape while the metal is red-hot. The metal passes through a series of rollers, each a little closer to the next one. As the steel passes through the last rollers, hot water is sprayed over it, forming a bluish scale. This steel is fairly uniform in quality and is used for many different kinds of parts. Hot-rolled bars of the best quality are used to produce cold-finished steels. Cold-finished steels are used when great accuracy, better surface finish, and certain mechanical properties are needed. There are several ways of producing cold-finished bars. The most common results in what is called cold-worked steel. After the scale from the hot-rolled bars is removed, one of two techniques is employed: (1) The bars are cold-drawn, that is, drawn through dies a few thousandths smaller than the original bar. (2) The steel is cold-rolled, that is, rolled cold to the exact size.

Drill rod   Drill rod is a grade of high-carbon steel or high-speed steel. It is finished by grinding and polishing so that the outside is smooth and very accurate in size. You can identify drill rod by its shiny surface, which is much smoother than any of the other steels used in the shop. Drill-rod bars are generally made in 3-foot lengths and come in round, hexagonal, and square shapes. Drill rod is more expensive than most other steels.

Cast iron   Cast iron is used for the heavy parts of many machines. It is the most common material for making castings. Cast iron is low in cost and wears well. It is very brittle, however, and cannot be hammered or formed. It contains 2 to 4 percent carbon. The basic kinds of cast iron are white iron, gray iron, and malleable iron. Malleable iron is a particular kind of cast iron, made more malleable by an annealing procedure. Malleable-iron castings are not so brittle or hard. They can stand a great deal of hammering. Many plumbing fixtures are made of malleable iron. Nodular iron is a kind of cast iron that is even better for withstanding shocks, blows, and jerks.

(1271 words)

From *Introduction to Basic Manufacturing Processes and Workshop Technology*
by Singh, R.

New Words and Expressions

annealing [ə'niːliŋ] n.	退火
casting n. & v.	铸造;铸件 & 浇铸
chisel ['tʃizəl] n.	凿子
chromium ['krəumiəm] n.	铬
cutting-edge n.	切削刃,刃口
detrimental [ˌdetri'mentl] a.	不利的,有害的
finish n. & v.	光洁度 & 精加工
fixture n.	固定装置,设备
forming n.	成型,成型加工
grind [graind] v.	磨削,研磨
hardening n.	硬化,淬火
hexagonal [hek'sægənl] a.	六边的,六角形的
impart [im'pɑːt] v.	给予,赋予
impurity n.	不纯,杂质
jerk [dʒəːk] n.	急拉,急推
manganese ['mæŋgəniːz] n.	锰
metallurgist [mə'tælədʒist] n.	冶金家
molybdenum [mə'libdənəm] n.	钼
phosphorus ['fɔsfərəs] n.	磷
polish v.	抛光
roller n.	轧辊;滚轴

scale $v.$	生成氧化皮
sulphur ['sʌlfə] $n.$	硫黄
susceptible [sə'septəbl] $a.$	易受影响的
trace $n.$	痕量
tungsten ['tʌŋstən] $n.$	钨
vanadium [və'neidiəm] $n.$	钒
warp ['wɔːp] $v.$	弯曲,变形
alloying agent	合金添加剂
bar steel	条钢,棒钢
drill rod	钻杆
ferrous metal	黑色金属
hardness penetration	淬硬深度,淬火深度,硬化深度
heat treatment	热处理
hot working	热加工,热作
nodular ['nɔdʒələ] iron	球墨铸铁
nonferrous metal	有色金属
plain carbon steel	普通碳钢,碳素钢
rod steel	圆钢,钢杆
rolled steel	轧钢
structural steel	结构钢
ultimate strength	极限强度
wrought [rɔːt] iron	锻铁;熟铁
yield point	屈服点

📖 Text Understanding

Ⅰ. Decide whether the following statements are true (T) or false (F) according to the passage.

1. Nonferrous metals should not contain any iron.
2. Besides iron and carbon, plain carbon steels usually contain a small amount of silicon, tungsten, phosphorus, and manganese.

3. Medium-carbon steels are used for forgings and other applications where increased strength and a certain amount of ductility are necessary.
4. Nickel is one of the major constituents of stainless steels.
5. Although phosphorus has an undesirable effect on steel, less than 0.5 percent of phosphorus increases tensile strength.
6. Drill rod bars are readily to be identified because of their shiny surface.

II. Fill in the blanks with proper words according to the passage.

1. Metals are divided into ferrous and nonferrous types according to whether they contain _____ or not.
2. Steel is an alloy containing chiefly _____, _____, and certain other elements in varying amounts.
3. High-carbon steels are used for high-strength parts such as springs, _____, and _____.
4. Manganese possesses the property of aiding in increasing the depth of _____ _____.
5. Sulphur is generally regarded as detrimental to _____ of steel and to _____ of steel treated to high tensile strength.
6. Nickel steels are less susceptible to _____ and _____ than most other steels.
7. Rolled steels include _____ steels, _____ steels, and _____ steels.
8. The basic kinds of cast iron are _____, _____, and _____.

III. Give brief answers to the following questions.

1. What is the classification of engineering materials according to the passage?

2. Why is carbon the most important alloying element in steel? What's the relationship between the carbon content, the tensile strength, and the hardness of the steel?

3. What is the definition of alloy steels? How many different kinds of alloy steels are used in a car?

Vocabulary Building

IV. Fill in the table below by giving the corresponding translation.

English	Chinese
	退火
	痕量
finish	
tungsten	
	淬硬深度
	轧钢
drill rod	
ferrous metal	
	合金添加剂

V. Match the items listed in the following two columns.

1. annealing
2. hardness penetration
3. die
4. stainless steel
5. fatigue strength
6. grinding

a. a device used for shaping metal
b. steel containing nickel that makes it resistant to corrosion
c. a heat treatment process used to soften previously cold-worked metal
d. the value of stress at which failure occurs after a number of cycles
e. removing significant portions of metal from the blade
f. the depth up to which a material is hardened after putting through a heat treatment process

Translation Practice

VI. Translate the following paragraph into English.

从特点来看,金属不透明,有延展性,是热和电的良导体。金属分为黑色金属和有色金属。前者含铁,后者不含铁。某些元素加进钢里能够改善钢的性能。例如:加铬后能耐腐蚀;加钨后可增加硬度。铝、铜、合金、黄铜等是常用的有色金属。

Section C
Nonmetallic Materials

Nonmetallic is a broad category. It comprises organic materials of natural origin like wood, leather, and natural rubber. It includes materials such as plastics and paper which are manufactured at least in part from natural substances. It also includes inorganic (mineral) materials like glass, ceramics, and concrete. Fig. 7-11 illustrates the generic relationship of common nonmetallic materials.

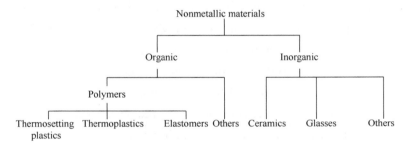

Fig. 7-11   Generic relationship of common nonmetallic materials

General Properties of Nonmetallic Materials

Nonmetallic materials have varied properties, and few characteristics are applicable to all of them. Two that are almost universal are low electrical and heat conductivity. With the exception of carbon, nonmetallic materials, when dry, are nonconductors.

Nonmetallic materials are usually less tough and less strong than metals, except that inorganic materials normally have very high compressive strengths. The inorganic materials also have superior high-temperature properties. Resistance to corrosion is a common property with many nonmetallics. Ease of fabrication is a property shared by polymers, wood, and some other organic materials.

Polymers: Plastics and Elastomers

A plastic is an organic polymer available in some resin form or a form derived from the basic polymerized resin. These forms include liquids and pastes for embedding, coating, and adhesive bonding. They also encompass molded, laminated, or formed shapes including sheet, film, and larger bulk shapes. While there are numerous minor classifications for polymers, depending upon how one wishes to categorize them, nearly all can be placed in one of two major classifications. These two major plastic-material classes are thermosetting materials (or thermosets) and thermoplastic materials, as shown in Fig. 7-11. Although Fig. 7-11 shows elastomers separately (and they are a separate application group), elastomers, too, are either thermoplastic or thermosetting, depending on their chemical nature.

As the name implies, thermosetting plastics, or thermosets, are cured, set, or hardened into a permanent shape. This curing is an irreversible chemical reaction known as cross-linking, which usually occurs under heat. For some thermosetting materials, however, curing is initiated or completed at room temperature. Even then, however, it is often the heat of the reaction, or the exotherm, which actually cures the plastic material. Such is the case, for instance, with room-temperature-curing epoxy, polyester, or urethane compounds.

Thermoplastics differ from thermosets in that they do not cure or set under heat. They merely soften, when heated, to a flowable state in which under pressure they can be forced or transferred from a heated cavity into a cool mold. Upon cooling in a mold, thermoplastics harden and take the shape of the mold. Since thermoplastics do not cure or set, they can be remelted and rehardened by cooling many times. Thermal aging, brought about by repeated exposure to the high temperatures required for melting, causes eventual degradation of the material, and so limits the number of reheat cycles.

The term "elastomers" includes the complete spectrum of elastic or rubberlike polymers which are sometimes randomly referred to as rubbers, synthetic rubbers, or elastomers. More properly, however, rubbers are natural materials, and synthetic rubbers are polymers which have been synthesized to reproduce consistently the best properties of natural rubber. Since such a large number of rubberlike polymers exist, the broad term "elastomer" is most fitting and most commonly used.

Other Organic Materials

Carbon and graphite Carbon is a very common element and the key constituent of all organic materials. In an uncombined pure state, it exists as diamond or graphite. In a less pure state, it exists as charcoal, coal, or coke (amorphous carbon). Both amorphous carbon and graphite are produced in structural shapes when the particles are bonded together with elemental carbon.

Carbon and graphite exhibit properties similar to those of ceramics with two major exceptions. They are electrically and thermally conductive. The ceramiclike properties include greater compressive than tensile strength, a lack of malleability and ductility, and a resistance to high temperatures and corrosive environments. The usable temperature limits for carbon and graphite are on the order of 2,400℃ and even higher. Strength is actually higher at elevated than at room temperatures. Specific electrical resistance ranges from a low of 0.004Ω-in (graphite) to 0.0022Ω-in (carbon).

The production of carbon and graphite components involves two processes: (1)

molding or extrusion followed by oven baking; (2) machining. High pressures and consequently significant die or mold costs are involved in the first method, which therefore is economically advantageous only for large-quantity production. Machining is more suitable for limited or moderate quantities.

Carbon and graphite components have mechanical, metallurgical, chemical, electrical, and nuclear applications. Typical uses include electrodes for the production of metals and chemicals in electric-arc furnaces, lighting electrodes, brushes for electric motors, electrodes in electrolytic cells, crucibles, molds for metal casting, resistance-furnace parts, and rocket components when high-temperature and thermal-shock resistance are important.

Carbon and graphite are available in round, square, and rectangular section. Round bars commonly stocked range from 3 to 1,100mm in diameter and from 300 to 2,800mm in length. Rectangular bars range from 13 by 100 by 400mm long to 600 by 750mm by 4.5m long.

Graphite is favored over carbon for applications requiring extensive machining. Graphite machines fairly well, and tolerances comparable with those of rough machining of metals can be achieved. Carbon members are recommended if only cutoff or other minimum machining is required.

Wood Wood has a number of desirable properties. It machines and fastens easily; it is attractive, as it has a high strength-to-weight ratio; when dry, it has good electrical-, heat-, and noise-insulating properties; it is long-lasting in dry environments; and it accepts preservative treatment readily. On the negative side are its directional-strength characteristics (because of its grain), its large dimensional change and tendency to warp with changes in moisture content, its susceptibility to rot in moist environments, and its poor abrasion resistance.

Ceramics and Glasses

Ceramics and glasses are nonorganic, nonmetallic materials made by fusing clays and other "earthy" materials which usually contain silicon and oxygen in various compositions with other materials.

Ceramics are hard, strong, brittle, and heat- and corrosion-resistant and are electrical insulators. They are used when these properties are important, particularly heat and corrosion resistance and electrical nonconductivity. Glass is used when transparency is important in addition to these properties.

(1036 words)

From *Introduction to Basic Manufacturing Processes and Workshop Technology*
by Singh, R.

New Words and Expressions

amorphous [əˈmɔːfəs] a.	无定形的，无组织的
cavity [ˈkæviti] n.	空腔；沟槽
ceramics [səˈræmiks] n.	陶瓷
charcoal [ˈtʃɑːkəul] n.	木炭；活性炭
coke [kəuk] n.	焦炭
compound n.	化合物，混合物
cross-linking n.	交联
crucible [kruːsibl] n.	坩埚
cure v.	使硬化，凝固
elastomer [iˈlæstəmə] n.	弹性体，弹性材料
embed v.	嵌入，内嵌
encompass [inˈkʌmpəs] v.	围绕，包围
epoxy [iˈpɔksi] n.	环氧树脂
exotherm [ˈeksəuˌθəːm] n.	温升，放热
extrusion [iksˈtruːʒn] n.	挤压加工
generic [dʒəˈnerik] a.	类属的；通用的
grain n.	晶粒，纹理
graphite [ˈgræfait] n.	石墨
metallurgical [ˌmetəˈləːdʒikl] a.	冶金的，冶金学的
polyester [ˌpɔliˈestə] n.	聚酯纤维
polymer [ˈpɔlimə] n.	聚合物，聚合体
polymerize [ˈpɔliməraiz] v.	使聚合
rectangular [rekˈtæŋgjələ] a.	矩形的，长方形的
resin [ˈrezin] n.	树脂
set v.	凝固，凝结
spectrum [ˈspektrəm] n.	范围；色谱，光谱
susceptibility [səˌseptəˈbiləti] n.	敏感性，灵敏度
tolerance n.	公差
transparency [trænsˈpærənsi] n.	透明，透明度
urethane [ˈjurəθein] n.	尿烷
adhesive bonding	粘接，黏合剂

electric-arc furnace	电弧炉
elemental carbon	元素碳
preservative treatment	防腐处理
synthetic [sin'θetik] rubber	合成橡胶
thermal aging	热老化
thermoplastic material	热塑性材料
thermosetting ['θəːməusetiŋ] material	热固性材料

Exercises

📖 Text Understanding

Ⅰ. Choose the best answer according to the passage.

1. In the following four choices, _____ has no association with natural substances.
   A. leather   B. plastics   C. paper   D. ceramics
2. Nonmetallic materials have some common properties shared by most nonmetallics except _____.
   A. low electrical conductivity   B. low heat conductivity
   C. resistance to corrosion   D. ease of fabrication
3. Depending on their chemical nature, elastomers are _____.
   A. thermoplastic   B. thermosetting
   C. either A or B   D. neither A nor B
4. _____ causes eventual degradation of the material and so limits the number of reheat cycles.
   A. Thermal aging   B. Fatigue strength
   C. Thermal circulation   D. Cross-linking
5. In a less pure state, carbon couldn't exist as _____.
   A. charcoal   B. coal   C. coke   D. diamond
6. Carbon and graphite are available in the following sections except in _____ section.
   A. rectangular   B. triangle   C. round   D. square

Ⅱ. Fill in the blanks with proper words according to the passage.

1. _____ is a property shared by polymers, wood, and some other organic materials.

2. The two major plastic-material classes are _____ materials and _____ materials.
3. Thermoplastics differ from thermosets in that _____ under heat.
4. In an uncombined pure state, carbon exists as _____ or _____.
5. Carbon and graphite exhibit properties similar to those of ceramics with two major exceptions: they are _____ conductive and _____ conductive.
6. Graphite is favored over carbon for applications _____.
7. Because of its grain, wood has _____ characteristics.
8. Ceramics and glasses are nonorganic and nonmetallic materials made by _____ and other _____ materials.

III. Give brief answers to the following questions.

1. What is the generic relationship of common nonmetallic materials?

2. What's the difference between rubbers and synthetic rubbers?

3. What are the two processes of the production of carbon and graphite components? Which one is economically preferred to mass production?

IV. What are the differences between thermosets and thermoplastics? Explain the differences according to the information offered in the following table.

Information	Thermosets	Thermoplastics
when heated		
characteristics		
in fabrication		

Vocabulary Building

V. Fill in the blanks with the words derived from the roots given in the following sentences.

1. Carbon and graphite components have mechanical, _____ (metallurgy), chemical, electrical, and nuclear applications.
2. The basic kinds of cast iron are white iron, gray iron, and _____ (malleability) iron.

3. Nickel steels are less _____ (susceptibility) to warping and scaling than most other steels.
4. The melting temperatures and the _____ (crystal) temperatures have a great effect on metals and alloys.
5. Cross-linking is a(n) _____ (reverse) chemical reaction which occurs only once.

✎Translation Practice

Ⅵ. Translate the following paragraphs from the passage into Chinese.

1. The term "elastomers" includes the complete spectrum of elastic or rubberlike polymers which are sometimes randomly referred to as rubbers, synthetic rubbers, or elastomers. More properly, however, rubbers are natural materials, and synthetic rubbers are polymers which have been synthesized to reproduce consistently the best properties of natural rubber. Since such a large number of rubberlike polymers exist, the broad term "elastomer" is most fitting and most commonly used.

2. Graphite is favored over carbon for applications requiring extensive machining. Graphite machines fairly well, and tolerances comparable with those of rough machining of metals can be achieved. Carbon members are recommended if only cutoff or other minimum machining is required.

💬 Translation Skill Ⅶ

句法的处理(1)—— 被动语态的翻译

科技英语在句法方面最突出的特点就是大量使用被动语态。著名语言学家 Quirk 曾说过,科技英语中被动语态的使用率要比主动语态高出十倍。根据英国利兹大学的统计,科技英语中的谓语至少三分之一是被动语态。由此可见,被动语态在科技英语中的地位之重要,使用之频繁。究其原因主要有二:一是被动语态比主动语态主观色彩更少,更能客观地反映事实;二是被动语态比主动语态更能说明需要论证的对象,因为在被动句中,所需论证、说明的对象充当句子的主语,其位置鲜明、突出,更能引起读者的注意。

因此,在科技英语翻译成汉语的过程中,不能过分受限于英语中的被动语态,而要根据汉语的语言习惯与特点,灵活处理被动结构,从而使译文更加准确、流畅。下面,我们将从三方面介绍被动语态的翻译技巧。

1. 译成汉语主动句

在科技英语中,被动语态的使用很常见,但中文中的被动语态相对较少,所以在汉译时,可以采取语态转换,将英语中被动语态的句子转译为中文主动语态的句子,这样更加符合中文的语言特点和习惯。

A. 主语不变,译为形式主动句

<u>Nuclear power's danger</u> to health, safety, and even life itself <u>can be summed up</u> in one word: radiation.

译文:<u>核能对健康、安全、甚至对生命本身构成的危害可以概括</u>为一个词——辐射。

<u>Two different lifetime limiting effects have been addressed</u>, namely creep rupture and low cycle fatigue.

译文:<u>两种不同的限制寿命的影响已经解决</u>,即蠕变断裂和低周期疲劳。

B. 主语译作宾语,介词后的名词译作主语,谓语被动变主动

A right kind of fuel <u>is needed</u> for <u>an atomic reactor</u>.

译文:<u>原子反应堆需要</u>一种合适的燃料。

Thermal noise <u>is caused</u> by the thermal motion of electrons in all conductors.

译文:所有导体中电子的热运动都会产生热噪声。

C. 增加主语,原句主语译作宾语,谓语被动变主动

<u>These ideas are developed</u> further, in the context of a digital communication system.

译文:<u>后文将以数字通信系统的角度进一步阐述这些思想</u>。

Silver <u>is known to</u> be better than copper in conductivity.

译文:<u>大家知道银的导电性比铜好</u>。

2. 译成汉语被动句

虽然被动语态在汉语中的使用较少,但英译汉时也会出现保留英语中的被动语态的情况。通常来说,英语被动句中强调被动动作或动作实施者时,会翻译成汉语的被动句。而且现在受外来语的影响,汉语中被动语态的使用也有一定上升趋势。

A. 译作"被"字

<u>Columns are defined as</u> members that carry loads chiefly in compression.

译文:<u>立柱被定义为</u>主要承载压力的构件。

Elevator drive system <u>is mounted at</u> the bottom of the building.

译文:电梯驱动系统被安装在建筑的底部。

B. 译作"隐性被动语态",用其他词代替"被"来表达被动意义,如"受"、"由"、"给"、"遭"、"使"等

These signals <u>are produced by</u> colliding stars or nuclear reaction in outer space.

译文:这些讯号<u>是由</u>外层空间的星球碰撞或者核反应<u>所造成</u>的。

Vehicle dynamics <u>is influenced by</u> several uncertain conditions (weather, road, state, etc.) and perturbations (aerodynamic drag forces, undesirable yaw moments, etc.).

译文：车辆动力学<u>受到</u>几个不确定条件(天气、道路、状态等)以及扰动因素(空气动力阻力、不利的横摆力矩等)<u>的影响</u>。

3. 译成汉语无主句

汉语中常常会出现省略主语的现象，所以在翻译英语被动语态时，也有译为汉语无主句的情况。

<u>It is reported that</u> no maintenance was applied on the shaft during operation life.

译文：<u>据称</u>，该驱动轴在工作年限内不需要进行维修。

<u>Stress must be laid</u> on the development of the electronics industry.

译文：<u>必须强调</u>电子工业的发展。

总之，英语与中文作为两种形态的语言，具有各自不同的语言特点。在翻译时，应该充分考虑到各自的行文习惯，从而最好地实现两者的转化衔接。通常来说，在科技英语中，被动语态大多处理为汉语主动句、无主句，在较少情况下处理为汉语被动句。需要指出的是，被动语态的翻译并不只有这几种固定的译法，在翻译时，译者需灵活变通，综合考虑英语的语言特点，以便译出最流畅、最符合中文习惯的译文。

Unit 8

Metal Working Processes

Before-Class Reading

Answer the following question before reading the passage:

1. Have you any ideas what heat treatment is? If you have, describe one process in detail; if not, consult relevant information on the internet to get some general ideas before reading Section A.

Section A
Heat Treatment

Heat treatment is a process utilized to change certain characteristics of metals and alloys in order to make them more suitable for a particular kind of application. In general, heat treatment is the term for any process employed which changes the physical properties of a metal by either heating or cooling.

When properly performed, heat treatment can greatly influence mechanical properties such as strength, hardness, ductility, toughness, and wear resistance. The various heat treatment processes appropriate to steels are:

- Hardening
- Tempering
- Annealing
- Normalizing
- Case hardening

Hardening Hardening is a process of heating and cooling steel to increase its hardness and tensile strength, to reduce its ductility, and to obtain a fine grain structure. The procedure includes heating the metal above its critical point or temperature, followed by rapid cooling. As steel is heated, a physical and chemical change takes place between the iron and carbon. The critical point, or the critical temperature, is the point at which the steel has the most desirable characteristics. When steel reaches this temperature (1400°F-1600°F), the change is ideal to make for a hard, strong material if it is cooled quickly. If the metal cools slowly, it changes back to its original state. By plunging the hot metal into water, oil, or brine (quenching), the desirable characteristics are retained. The metal is very hard and strong and less ductile than before.

The hardening procedures are:

1. Light the furnace, and allow it to come to the right temperature.

2. Place the metal in the furnace, and heat it to the critical temperature.

3. Select the correct cooling solution. Some steels can be cooled in water, and others must be cooled in oil or brine. Water is the most widely used material for quenching carbon steels because it is inexpensive and effective. Brine is usually made by adding about 9% of common salt to the water. Brine helps to produce a more uniform hardness. Oil is used for a somewhat slower speed of quenching. Most oils used for quenching are mineral oils.

4. Remove the hot metal with tongs and plunge it into the cooling solution.

5. A properly hardened piece of steel will be hard and brittle and have high tensile strength. It will also have internal strains. If left in this state, these internal strains could cause the metal to crack.

Tempering Tempering is a process of reducing the degree of hardness and strength and increasing the toughness. It removes the brittleness from a hardened piece. It is a process that follows the hardening procedures and makes the metal as hard and tough as possible. Tempering is done by reheating the metal to low or moderate temperature, followed by quenching or by cooling in air. This temperature will remove internal stress setup during quenching, remove some or all of the hardness, and increase the toughness of the material. As the metal is heated for tempering, it changes in color. These colors are called temper colors.

Annealing Annealing is a heat treating process used to soften previously cold-worked metal by allowing it to recrystallize. The term annealing refers to a heat treatment in which a material is exposed to an elevated temperature for an extended time period and then slowly cooled. Ordinarily, annealing is carried out to (1) relieve stresses; (2) increase softness, ductility, and toughness; and/or (3) produce a desired microstructure. A variety of annealing heat treatments are possible, such as stress-relief annealing, spheroidized annealing, and full annealing.

Any annealing process consists of three stages and time is an important parameter in these procedures:

1. Heating to the desired temperature;

2. Holding or "soaking" at the temperature;

3. Slowly cooling, usually to room temperature.

Annealing process is a heat treatment that is used to negate the effects of cold work that is to soften and increase the ductility of a previously strain-hardened metal.

Normalizing　　Normalizing process resembles full annealing except that, whilst in annealing the cooling rate is deliberately retarded, in normalizing the cooling rate is accelerated by taking the work from the furnace and allowing it to cool in free air. Provision must be made for the free circulation of cool air, but draughts must be avoided.

In the normalizing process, it can be seen that the steel is heated to approximately 50℃ above the upper critical temperature line. This ensures that the transformation to fine grain austenite corrects any grain growth or grain distortion that may have occurred previously. Again, the steel is cooled in free air and the austenite transforms into fine grain pearlite and cementite. The fine grain structure resulting from the more rapid cooling associated with normalizing gives improved strength and toughness to the steel but reduces its ductility and malleability. The increased hardness and reduced ductility allows a better surface finish to be achieved when machining.

Case hardening　　Sometimes special characteristics are required in metal such as hard outer surface and soft, tough, and more strength oriented core or inner structure of metal. This can be obtained by case hardening process. It is the process of carburization, i.e. saturating the surface layer of steel with carbon or some other substances by which the outer case of the object is hardened whereas the core remains soft. It is applied to very low carbon steel. It is performed for obtaining hard and wear resistance on surface of metal and higher mechanical properties with higher fatigue strength and toughness in the core.

This process can be done on such items as hammer heads, piston pins, and other items that must stand a good deal of shock and wear. It can never be used on anything that must be sharpened by grinding.

(946 words)

https://learnmechanical.com/heat-treatment/

New Words and Expressions

austenite [ˈɔstəˌnait] n.　　　　　　　　　　　奥氏体

brine [ˈbrain] n.　　　　　　　　　　　　　　盐水

carburization [ˌkɑːbjuraiˈzeiʃən] n.	表面渗碳法
cementite [siˈmentait] n.	渗碳体；碳化铁
draught [drɑːft] n.	气流
normalizing [ˈnɔːməlaiziŋ] n.	正火
pearlite [ˈpəːlait] n.	珠光体
quenching [ˈkwentʃiŋ] n.	淬火；骤冷
recrystallize [riːˈkristəlaiz] v.	再结晶，重结晶
saturate [ˈsætʃəreit] v.	浸透；使饱和
tempering [ˈtempəriŋ] n.	回火
tong [ˈtɔŋ] n.	钳子
case hardening	表面硬化，表面淬火
critical point	临界点
critical temperature	临界温度
fine grain	细晶粒
full annealing	完全退火
hammer head	锤头，榔头
internal strain	内应变
internal stress	内应力
negate [niˈgeit] v.	否定，取消
piston pin	活塞销
spheroidized [ˌsfiərɔiˈdaizd] annealing	球化退火
stress-relief annealing	消除应力退火，去应力退火

Exercises

📖 Text Understanding

Ⅰ. Decide whether the following statements are true (T) or false (F) according to the passage.

1. The hardening operation consists of heating the steel above the critical temperature and then slowly cooling it in some suitable solution.
2. Hardening is the process of softening steel to relieve internal strain.
3. Hardening procedure follows tempering and makes the metal as hard and tough as

possible.

4. Brine is the most widely used material for quenching carbon steels because it functions more quickly.
5. In normalizing, the cooling rate is accelerated by taking the work from the furnace and allowing it to cool in the draughts.
6. Case hardening can never be used on anything that must be sharpened by grinding.
7. Tempering process consists of reheating the metal to low or moderate temperature, followed by quenching or by cooling in air.

II. Match the term in the left column with the process in the right column.

1. annealing   a. It is the process of saturating the surface layer of steel with carbon or some other substances by which the outer case of the object is hardened whereas the core remains soft.

2. case hardening   b. The procedure includes heating the metal above its critical point or temperature, followed by rapid cooling.

3. hardening   c. It resembles full annealing, but the cooling rate is accelerated by taking the work from the furnace and allowing it to cool in free air.

4. normalizing   d. It is done by reheating the metal to low or moderate temperature, followed by quenching or by cooling in air.

5. tempering   e. It is a heat treatment in which a material is exposed to an elevated temperature for an extended time period and then slowly cooled.

III. Give brief answers to the following questions.

1. What is heat treatment?

2. What is the critical temperature for steel, and what is the ideal temperature for its hardening process?

3. What is the most obvious drawback of hardening? How to solve the problem?

4. Give at least three examples of different kinds of annealing.

✎Translation Practice

IV. Translate the following paragraph into Chinese.

低碳钢由于含碳量低,因此在经受这种热处理时材质不可能变硬。若欲在低碳钢制成的零件表面获得硬的表面层,就须进行表面硬化处理。氰化(cyaniding)是一种表面硬化方法。氰化时,将工件置入氰化钠的溶池中5～30分钟。工件经过这一处理后,再将其淬入水或油中,于是形成了厚度为0.254～0.381mm 的十分硬的表面层。

In-Class Reading

Section B
Forming

Forming can be defined as a process in which the desired size and shape are obtained through the plastic deformation of a material. The stresses induced during the process are greater than the yield strength, but less than the fracture strength of the material. The type of load may be tensile, compressive, bending, or shearing, or a combination of these. This is a very economical process as the desired shape, size, and finish can be obtained without any significant loss of material. Moreover, a part of the input energy is fruitfully utilized in improving the strength of the product through strain hardening.

The forming processes can be grouped under two broad categories, namely cold forming and hot forming. If the working temperature is higher than the recrystallization temperature of the material, then the process is called hot forming. Otherwise the process is termed as cold forming. The flow stress behavior of a material is entirely different above and below its recrystallization temperature. During hot working, a large amount of plastic deformation can be imparted without significant strain hardening. This is important because a large amount of strain hardening renders the material brittle. The frictional characteristics of the two forming processes are also entirely different. For example, the coefficient of friction in cold forming is generally of the order of 0.1, whereas that in hot forming can be as high as 0.6. Further, hot forming lowers down the material strength so that a machine with a reasonable capacity can be used even for a product having large dimensions.

The typical forming processes are rolling, forging, extrusion, drawing, and deep drawing. For a better understanding of the mechanics of various forming operations, we shall briefly discuss each of these processes.

Rolling

Rolling is the most rapid method of forming metal into desired shapes by plastic deformation through compressive stresses using two or more than two rolls. It is one of the most widely used metal working processes. The main objective of rolling is to convert larger sections such as ingots into smaller sections which can be used either directly as rolled state or as stock for working through other processes. The coarse structure of cast ingot is convened into a fine grained structure using rolling process as shown in Fig. 8-1. Significant improvement is accomplished in rolled parts in their various mechanical properties such as toughness, ductility, strength, and shock resistance. The majority of steel products are converted from the ingot form by the process of rolling. To the steel supplied in the ingot form, the preliminary treatment imparted is the reduction in its section by rolling. As shown in Fig. 8-1, the crystals in parts are elongated in the direction of rolling, and they start to reform after leaving the zone of stress. Hot rolling process is widely used in the production of large number of useful products such as rails, sheets, structural sections, plates, etc.

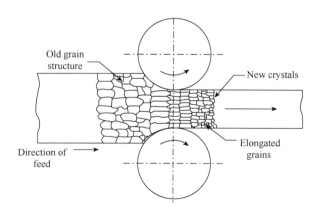

Fig. 8-1 Grain refinement in hot rolling process

Forging

In forging, the material is squeezed between two or more dies to alter its shape and size. Depending upon the complexity of the part, forging is carried out as open die forging and closed die forging. In open die forging (Fig. 8-2 (a)), the metal is compressed by repeated blows from a mechanical hammer and the shape is manipulated manually. In closed die forging (Fig. 8-2 (b)), the desired configuration is obtained by squeezing the workpiece between two shaped and closed dies.

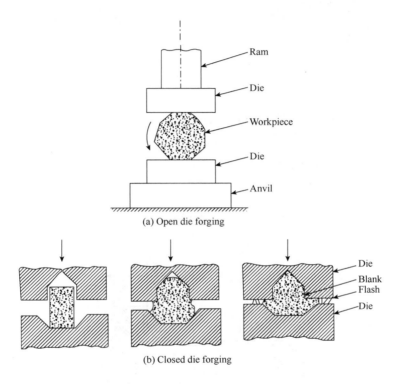

Fig. 8-2 Different types of forging

Extrusion

Extrusion (Fig. 8-3) is a process basically similar to the closed die forging. But in this operation, the workpiece is compressed in a closed space, forcing the material to flow out through a suitable opening, called a die. In this process, only the shapes with constant cross-sections (die outlet cross-section) can be produced.

Drawing

Drawing is the pulling of metal through a die or a set of dies for achieving a reduction in a diameter. The material to be drawn is reduced in diameter. Fig. 8-4 represents the operation schematically. When high reduction is required, it may be necessary to perform the operation in several passes. Large quantities of wires, rods, tubes, and other sections are produced by drawing process which is basically a cold working process.

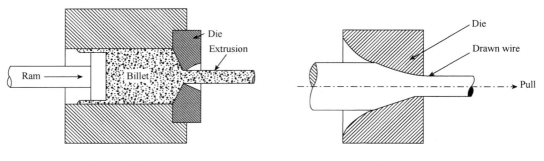

Fig. 8-3 Extrusion   Fig. 8-4 Drawing

Deep Drawing

In deep drawing, cylindrical shaped parts such as cups, shells, etc. are obtained from a flat sheet metal with the help of a punch and a die. Fig. 8-5 shows the operation schematically. The sheet metal is held over the die by means of a blank holder to avoid defects in the product.

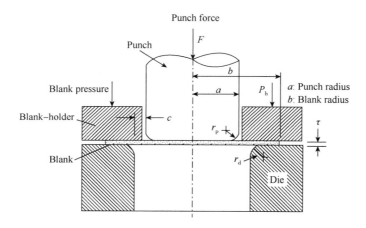

Fig. 8-5 Deep drawing

Advantages and Disadvantages of Hot and Cold Forming

Now that we have covered the various types of metal working operations, it would be appropriate that we provide an overall evaluation of the hot and cold working processes. Such a discussion will help in choosing the proper working conditions for a given situation.

During hot working, a proper control of the grain size is possible since active grain growth takes place in the range of the working temperature. As a result, there is no strain hardening, and therefore there is no need of expensive and time-consuming intermediate annealing. Of course, strain hardening is advisable during some operations (viz., drawing) to achieve an improved strength; in such cases, hot working is less advantageous. Apart from this, strain hardening may be essential for a successful completion of some processes (e.g., in deep drawing, strain hardening prevents the rupture of the material around the bottom circumstance where the stress is maximum). Large products and high strength materials can be worked upon under hot conditions since the elevated temperature lowers down the strength and, consequently, the work load. Moreover, for most materials, the ductility increase with temperature and, as a result, brittle materials can also be worked upon by the hot working operation. It should, however, be remembered that there are certain materials (viz., steels containing sulphur) which become more brittle at elevated temperature. When a very accurate dimensional control is required, hot working is not advised because of shrinkage and loss of surface metal due to scaling. Moreover, surface finish is poor due to oxide formation and scaling.

The major advantages of cold working are that it is economical, quicker, and easier to handle because here no extra arrangements for heating and handling are necessary. Further, the mechanical properties normally get improved during the process due to strain hardening. What's more, the control of grain flow directions adds to the strength characteristics of the product. However, apart from other limitations of cold working (viz., difficulty with high strength and brittle materials and large product sizes), the inability of the process to prevent the significant reduction brought about in corrosion resistance is an undesirable feature.

(1142 words)

From *Introduction to Basic Manufacturing Processes and Workshop Technology*
by Singh, R.

New Words and Expressions

anvil ['ænvil] *n.*	铁砧
blank *n.*	板坯,坯料
convene [kən'vi:n] *v.*	聚集,集合;召集
flash *n.*	飞边
ingot ['iŋgət] *n.*	铸块,锭
radius ['reidiəs] *n.*	半径
rail *n.*	钢轨,轨条
ram *n.*	冲头

sheet *n*.	薄钢板,薄板
shrinkage [ˈʃrɪŋkɪdʒ] *n*.	收缩
viz.	即,也就是
workpiece *n*.	工件
blank holder	压边圈,压料板;坯缘压牢器
cast ingot	铸锭
closed die forging	闭式模锻
coarse structure	粗糙结构,粗松组织
coefficient of friction	摩擦系数
deep drawing	深压成型,深拉
direction of feed	进刀方向,进给方向
intermediate annealing	中间退火
open die forging	开式模锻
strain hardening	应变硬化,加工硬化

Exercises

📖 Text Understanding

I. Fill in the blanks with proper words according to the passage.

1. The forming processes can be grouped under two broad categories, namely _____ and _____.

2. During hot working, a large amount of _____ can be imparted without significant strain hardening. This is important because a large amount of strain hardening renders the material _____.

3. The main objective of rolling is to convert larger sections such as ingots into smaller sections which can be used either directly in as _____ or as _____ for working through other processes.

4. In _____ die forging, the metal is compressed by repeated blows from a mechanical hammer and the shape is manipulated manually.

5. Extrusion is a process basically similar to _____.

6. When high reduction is required, it may be necessary to perform the operation in _____.

II. Match the term in the left column with the description in the right column.

1. deep drawing
2. drawing
3. extrusion
4. forging
5. rolling

a. Desired shapes are obtained by plastic deformation through compressive stresses using two or more than two rolls.

b. Cylindrical shaped parts such as cups, shells, etc. are obtained from a flat sheet metal with the help of a punch and a die.

c. The material is squeezed between two or more dies to alter its shape and size.

d. Only the shapes with constant cross-sections can be produced.

e. It is pulling of metal through a die or a set of dies for achieving a reduction in a diameter.

III. Decide whether the following statements are about hot forming (HF) or about cold forming (CF).

(　) 1. There is no strain hardening.
(　) 2. There is no intermediate annealing.
(　) 3. The mechanical properties get improved due to strain hardening.
(　) 4. It is economical, quicker, and easier to handle.
(　) 5. Brittle materials can be worked upon by the hot working operation.
(　) 6. The control of grain flow directions adds to the strength characteristics of the product.
(　) 7. A proper control of the grain size is possible.

IV. Give brief answers to the following questions.

1. What is forming? Give at least five examples of the typical forming processes.

2. What are the two broad categories of forming processes? How to define them respectively?

3. Why is hot working not advisable when an accurate dimensional control is needed?

Vocabulary Building

V. Fill in the table below by giving the corresponding translation.

English	Chinese
closed die forging	
intermediate annealing	
	飞边
	摩擦系数
cast ingot	
blank holder	
	拉拔
	挤压
	应变硬化

Translation Practice

VI. Translate the following sentences from the passage into Chinese.

1. Moreover, a part of the input energy is fruitfully utilized in improving the strength of the product through strain hardening.

2. In closed die forging, the desired configuration is obtained by squeezing the workpiece between two shaped and closed dies.

3. Large quantities of wires, rods, tubes, and other sections are produced by drawing process which is basically a cold working process.

4. The sheet metal is held over the die by means of a blank holder to avoid defects in the product.

5. However, apart from other limitations of cold working (viz., difficulty with high strength and brittle materials and large product sizes), the inability of the process to prevent the significant reduction brought about in corrosion resistance is an undesirable feature.

___

Section C
Casting

Casting is the introduction of molten metal into a cavity or mold where, upon solidification, it becomes an object whose shape is determined by mold configuration. Casting offers several advantages over other method of metal forming: it is adaptable to intricate shapes, to extremely large pieces, and to mass production; it can provide parts with uniform physical and mechanical properties throughout, and depending on the particular material being cast, the design of the part, and the quantity being produced, its economic advantages can surpass other processes.

Casting can be divided into two categories, namely ingot casting (which includes continuous casting) and casting to shape. Ingot castings are produced by pouring molten metal into a permanent or reusable mold. Following solidification, these ingots (or bars, slabs, or billets) are then further processed mechanically into many new shapes. Casting to shape involves pouring molten metal into molds in which the cavity provides the final useful shape, followed only by machining or welding for the specific application.

Ingot Casting

Ingot castings make up the majority of all metal castings and are separated into three categories: static cast ingots, semi-continuous or direct-chill cast ingots, and continuous cast ingots.

Static cast ingots Static ingot casting simply involves pouring molten metal into a permanent mold (Fig. 8-6). After solidification, the ingot is withdrawn from the mold and the mold can be reused. This method is used to produce millions of tons of steel annually.

Semi-continuous cast ingots A semi-continuous casting process is employed in the aluminum industry to

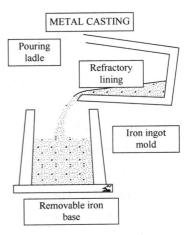

Fig. 8-6 Static ingot casting

produce most of the cast alloys from which rod, sheet, strip, and plate configurations are made. In this process, molten aluminum is transferred to a water-cooled permanent mold (Fig. 8-7 (a)) which has a movable base mounted on a long piston. After solidification, a solid "skin" is formed on the mold surface. The piston is moved down, and more metal continues to fill the reservoir (Fig. 8-7 (b)). Finally when the piston is moved its entire length, the process is stopped. Conventional practice in the aluminum industry utilizes suitably lubricated metal molds. However, technological advances have allowed major aluminum alloy producers to replace the metal mold (at least in part) by an electromagnetic field so that molten metal touches the metal mold only briefly, thereby making a product with a much smoother finish than that produced conventionally.

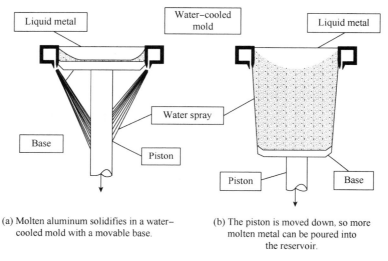

(a) Molten aluminum solidifies in a water-cooled mold with a movable base.

(b) The piston is moved down, so more molten metal can be poured into the reservoir.

Fig. 8-7　Semi-continuous casting

Continuous cast ingots　Continuous casting provides a major source of cast materials in the steel and copper industry and is growing rapidly in the aluminum industry. In this process molten metal is delivered to a permanent mold, and the casting begins much in the same way as in semi-continuous casting. However, instead of the process ceasing after a certain length of time, the solidified ingot is continually sheared or cut into lengths and removed during casting. Thus the process is continuous, the solidified bar or strip being removed as rapidly as it is being cast. This method has many economic advantages over the more conventional casting techniques. As a result, all modern steel mills produce continuous cast products.

Casting to Shape

Casting to shape is generally classified according to the molding process, the molding material, or the method of feeding the mold. There are four basic types of these casting processes: sand casting, permanent-mold casting, die casting, and centrifugal casting.

Sand casting This is the traditional method which still produces the largest volume of cast-to-shape pieces. It utilizes a mixture of sand grains, water, clay, and other materials to make high-quality molds for use with molten metal. This "green sand" mixture is compacted around a pattern (wood, plaster, or metal), usually by machines, to 20-80% of its bulk density. The basic components of a sand mold and of other molds as well are shown in Fig. 8-8. The two halves of the mold (the cope and the drag) are closed over cores necessary to form internal cavities, and the whole assembly is weighted or clamped to prevent floating of the cope when the metal is poured.

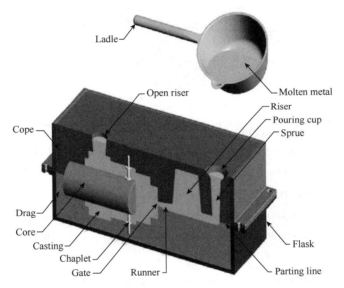

Fig. 8-8 Section through a sand mold showing the gating system and the risers

Other casting processes which utilize sand as a basic component are the shell, carbon dioxide, investment casting, ceramic molding, and plaster molding processes. In addition, there are a large number of chemically bonded sands which are becoming increasingly important.

Permanent-mold casting Many high-quality castings are obtained by pouring molten metal into a mold made of cast iron, steel, or bronze (Fig. 8-9).

Semipermanent mold materials such as aluminum, silicon carbide, and graphite may also be used. The mold cavity and the gating system are machined to the desired dimensions after the mold is cast: the smooth surface from machining thus gives a good surface finish and dimensional accuracy to the casting. To increase mold life and to make ejection of the casting easier, the surface of the mold cavity is usually coated with carbon soot or a refractory slurry. These also serve as heat barriers and control the rate of cooling of the casting. The process is used for cast iron and nonferrous alloys with advantages over sand casting, such as smoother surface finish, closer tolerance, and higher production rates.

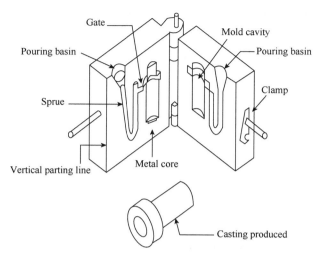

Fig. 8-9 A typical permanent mold

Die casting A further development of the permanent molding process is die casting. Molten metal is forced into a die cavity under pressures of 100-100,000 psi. Two basic types of die-casting machines are hot-chamber and cold-chamber. In the hot-chamber machine, a portion of the molten metal is forced into the cavity at pressures up to about 2,000 psi. The process is used for casting low-melting-point alloys such as lead, zinc, and tin.

In the cold-chamber process, the molten metal is ladled into the injection cylinder and forced into the cavity under pressures which are about 10 times as high as those in the hot-chamber process. High-melting-point alloys such as aluminum-, magnesium-, and copper-base alloys are used in this process. Die casting has the advantage of high production rate, high quality and strength, surface finish on the order of 40-100 microinch rms (root mean square), and close tolerances, with thin sections.

Rheocasting is the casting of a mixture of solid and liquid. In this process, the alloy to be cast is melted and then allowed to cool until it is about 50% solid and 50% liquid. Vigorous stirring promotes liquidlike properties of this mixture so that it can be injected in a die-casting operation. A major advantage of this type of casting process is expected to be much reduced die erosion due to the lower casting temperatures.

Centrifugal casting Inertial forces of rotation distribute molten metal into the mold cavities during centrifugal casting, of which there are three categories: true centrifugal casting, semicentrifugal casting, and centrifuging. The first two processes produce hollow cylindrical shapes and parts with rotational symmetry respectively. In the third process, the mold cavities are spun at a certain radius from the axis of rotation; the centrifugal force thus increase the pressure in the mold cavity.

The rotational speed in centrifugal casting is chosen to give between 40 and 60g acceleration. Dies may be made of forged steel or cast iron. Colloidal graphite is used on the dies to facilitate removal of the casting.

Successful operation of any metal-casting process requires careful consideration of mold design and metallurgical factors.

(1251 words)

From *Introduction to Basic Manufacturing Processes and Workshop Technology* by Singh, R.

New Words and Expressions

billet *n.*	方坯,坯锭
carbide [kɑːbaid] *n.*	碳化物
centrifuging *n.*	离心法
chaplet ['tʃæplət] *n.*	型芯撑
clamp *v.*	夹紧,固定住
continuous cast ingot	连续铸锭
cope [kəup] *n.*	上型箱,上砂箱
core *n.*	型芯
drag [dræg] *n.*	下型箱,下砂箱
feed *v.*	进刀,进料,进给
flask *n.*	砂箱,上箱
inertial [iˈnəːʃl] *a.*	惯性的
ladle [ˈleidl] *v.*	把……倒入,用长柄勺舀
plaster [ˈplɑːstə] *n.*	石膏;灰泥
rheocasting [ˈriːəukæstiŋ] *n.*	流变铸造
runner *n.*	流道,分流道
slab *n.*	板坯
slurry [ˈsləːri] *n.*	泥浆
solidification [səˌlidifiˈkeiʃn] *n.*	凝固,固化
sprue [spruː] *n.*	注入口,浇口
symmetry [ˈsimətri] *n.*	对称;相仿

bulk density	体积密度,松装密度,堆积密度
carbon soot	炭黑
casting to shape	铸型
centrifugal [ˌsentriˈfjuːɡl] casting	离心式铸造
colloidal [kəˈlɔidəl] graphite	石墨乳
die casting	压铸;拉模铸造
direct-chill cast ingot	直冷式铸锭
gating system	浇铸系统
ingot casting	铸锭,钢锭浇铸
internal cavity	内腔
investment casting	熔模铸造,精密铸造
open riser	明冒口
parting line	分型线,分模线
permanent-mold casting	永久型铸造,恒模铸造
pouring basin	外浇口,浇口杯
root mean square	均方根
sand casting	砂型铸造
semicentrifugal casting	半离心式铸造
semi-continuous cast ingot	半连续性铸锭
static cast ingot	静态铸锭
true centrifugal casting	真正的离心式铸造

Exercises

📖 Text Understanding

Ⅰ. Choose the best answer according to the passage.

1. Among the following four casting operations, the mold may not be the permanent or reusable one in _____.
   A. static cast ingots   B. continuous cast ingots
   C. sand casting   D. die casting

2. Major aluminum alloy producers replace the metal mold by _____ so that the product will have a much smoother finish.
   A. a sand mold   B. a plaster mold
   C. a plastic mold   D. an electromagnetic field

3. Casting to shape is generally classified according to the following factors except _____.
   A. the quality of the mold   B. the molding process
   C. the method of feeding the mold   D. the molding material
4. Compared with sand casting, _____ is not the advantage of semipermanent mold casting.
   A. close tolerance   B. high production
   C. intricate shape   D. smooth surface finish
5. Inertial forces of rotation distribute molten metal into the mold cavities during _____.
   A. centrifugal casting   B. die casting
   C. permanent-mold casting   D. rheocasting
6. Successful operation of any metal-casting process requires careful consideration of _____.
   A. mold design   B. metallurgical factors
   C. both A and B   D. neither A nor B

II. Match the metal casting operations in the left column with the casting products in the right column.

1. static ingot casting
2. semi-continuous casting
3. continuous casting
4. sand casting
5. permanent-mold casting
6. hot-chamber die casting
7. cold-chamber die casting
8. centrifugal casting

a. most of the cast alloys in aluminum industry
b. the largest volume of cast-to-shape pieces
c. hollow cylindrical shapes and parts
d. millions of tons of steel
e. low-melting-point alloys such as lead, zinc, and tin
f. high-quality castings
g. cast material in the steel and copper industry
h. high-melting-point alloys such as aluminum-, magnesium-, and copper-base alloys

III. Give brief answers to the following questions.

1. What is casting? What are the advantages of casting?

   _____

2. Why are continuous cast ingots more advantageous than semi-continuous cast ingots?

   _____

3. Why is carbon soot or refractory slurry coated on the surface of mold cavities?

4. What is rheocasting and what is the major advantage of rheocasting?

▶ Vocabulary Building

Ⅳ. Fill in the table below by giving the corresponding translation.

English	Chinese
casting to shape	
centrifuging	
	凝固
	对称
internal cavity	
parting line	
	体积密度,松装密度
	均方根
static cast ingot	

✎ Translation Practice

Ⅴ. Translate the following sentences from Chinese into English.

1. 传统的铸造工艺仍然是当今用于生产造型复杂、体积大小不一的零部件的重要方法之一。(adaptable to)

2. 熔模铸造工艺比较适用于生产专业模具的型芯与型腔。(applicable to)

3. 理论上讲,用同一模具生产的零件属于可互换零件。(identical)

4. 永久型铸造技术可以用来复制从小到大错综复杂的模式。(range from…to…)

Translation Skill VIII

句法的处理(2)—— 否定句的翻译

中西方不同的思维方式对汉英语言有着深刻的影响,形成了汉英语言的不同特点。具体就否定表达而言,英语中表示否定的词比汉语多,否定句式也种类繁多、形式灵活多样,而汉语否定词汇及否定句式相对固定、简单。因此,要想准确地翻译否定句式,首先必须明晰英语中常见的否定句式表达,在充分了解其异同的基础上才能做到有的放矢,精准翻译出英语否定句。具体而言,科技英语的否定句式主要有:完全否定、部分否定、双重否定、形式上否定意义上肯定、含蓄否定及转移否定等六大类。下面,我们将逐一举例进行分析。

1. 完全否定

现代英语的完全否定通常由 no, not, never, nobody, nothing, nowhere, neither, not at all 等否定词构成,否定的是整体。

Carbon dioxide does not burn, neither does it support burning.

译文:二氧化碳既不自燃,也不助燃。

Never is aluminum found free in nature.

译文:铝在自然界从不以游离状态存在。

2. 部分否定

在英语中,当含有全体意义的代词如 all, each, every, both, everyone 等用于否定式谓语的句子中的时候,对该全体意义的代词及其对应的名词进行的是部分否定。

Notice that not all mechanical energy is kinetic energy.

译文:要注意,并非所有的机械能都是动能。

Both the instruments are not precision ones.

译文:这两台仪器并非都是精密仪器。

3. 双重否定

双重否定即否定之否定,其否定意义相互抵消而取得肯定意义,即我们常说的:"双重否定等于肯定"。

Sodium is never found uncombined in nature.

译文:A.自然界中从未发现不处于化合状态的钠。(保留双重否定词)
　　　B.自然界中已发现的钠都处于化合状态。(肯定表达)
A body at rest will never move unless it is acted on by an external force.
译文:A. 如果没有外力作用,静止的物体就不会移动。(保留双重否定词)
　　　B. 静止的物体只有在外力作用下才会移动。(肯定表达)

4.形式上否定意义上肯定

英语中有些句子形式上是否定的,但是内容含义上却是肯定的,如 nothing but(只,仅仅),no more than(不过),no less than(多达,高达),cannot/couldn't+too(应该非常,应该特别)等。

An explosion is nothing more than a tremendously rapid burning.
译文:爆炸只不过是非常急速的燃烧。

Pesticides can cause a lot of harm so you can never be too careful when using pesticides.
译文:杀虫剂会造成很大危害,所以使用杀虫剂时应特别小心。

5.含蓄否定

含蓄否定指英语中有些词或短语不与否定词连用,并且从句子的形式上也看不出任何否定的迹象,但其意义却是否定的。含蓄否定主要通过词汇手段来实现,如使用形式上是肯定但意义上为否定的单词或短语:lack, fail, deny, miss, decline, deficiency, exclusion, without, beyond, against, little, few, poor, blind, last, too…too…, rather than, out of 等。

It is the last type of machine for such a job.
译文:这是最不适合这种作业的机器。

The role of clouds is poorly understood.
译文:人们对于云层的作用知之甚少。

6.转移否定

转移否定是指英语否定词出现在句子谓语部分,但其在语义上却是否定另一部分的现象。翻译这类句子时如果单纯依赖语法分析,则译文会出现逻辑问题,因此,转移否定的翻译应从语义出发,根据上下文正确理解句意。另外,某些动词如 think, consider, suppose, believe 等以否定形式引导从句,在翻译时否定往往转移至从句部分。

The motor did not stop because the electricity was off.
译文:电机停止运转,并非因为电源切断了。

They don't consider that pure science is more important than applied science.
译文:他们认为纯科学并不比应用科学更重要。

以上为英语六大类否定句的句式介绍及基本译法阐释。简言之,英语否定句种类繁多、形式灵活多变,在翻译时需充分考虑否定句句式类型及语义逻辑关系,以精准翻译出英语中的各类否定句。

Unit 9

Mechanical Engineering Applications

Before-Class Reading

Answer the following question before reading the passage:
1. Can you make some speculation about the changes in business sectors after the outbreak of Covid-19 pandemic? Please give some examples.

Section A
Most Popular Industrial Robotic Applications for 2021 and Projections

Automation solution providers encountered a major shift in the market in 2021. The Covid-19 pandemic created a demand for automation in various business sectors, including those outside industrial manufacturing. Statistics published by the Association for Advancing Automation (A3) show that North American robotic sales increased by 3.5% in 2020 and are continuing to see positive growth. The pandemic brought many challenges to businesses like supply shortages, new safety precautions, and a major employment gap. Companies had to redesign business models and supply chain management to develop lean production processes with higher output to overcome these challenges. With many products facing supply shortages, industries such as manufacturing, life science, food processing, retail, and delivery/fulfillment have looked to automation as the solution. This has led to the automation of new applications and advancements to common applications.

Sorting Applications

Sorting and organizing technologies have seen a sharp increase in sales due to the medical and food supply shortages resulting from the Covid-19 pandemic. The sudden changes in consumers' buying habits increased the value of automation for companies across industries. Automation for sorting applications offers a long-term solution for labor shortages, social distancing requirements, and supply chain overhaul for businesses. The increased robotic market revenue from sorting applications brought forth complex and innovative solutions such as vision-based sorting and cleanroom robotics. Companies are using vision and machine learning with industrial robots (Fig. 9-1) to execute sorting tasks that involve vision-based decisions. Robotic companies released cleanroom robots for environments requiring sanitation tasks such as preparing syringes for coating, placing pills into bottles and capping them, and handling food packaging. Sorting robots continue to gain visibility and expand capabilities for significant improvements to production operations and product output.

Fig. 9-1　Industrial robots

Quality Inspection Applications

Quality control is used to eliminate defects and ensure customers receive products that meet their expectations. This application is pushed across all manufacturing industries to improve product quality. Cutting-edge vision technology has opened the door for automation to take over and improve this task. Operations can now implement quality control in new areas, eliminating defects that are not visible to the naked eye. An example of this technology is spectral imaging which detects color spectrums that human workers cannot perceive. These cameras are used in fresh-produce inspections to detect areas that are spoiled or foreign objects within the goods. Errors resulting from inadequate quality inspections can be dangerous, costly, and detrimental to a company's image. Vision technology gives an extra layer of security in ensuring all products leaving production are of the highest quality to consumers.

Collaborative Applications

Collaborative robots have made a come-back after two years of underperformance due to factory shutdowns during the Covid-19 pandemic. As discussed previously in the sorting application section, the recovery phase for companies requires a quick and efficient solution for supply shortages and safety restrictions. Companies are implementing collaborative technology in strategic ways to overcome employment gaps, social distancing restrictions, and the immediate need for leaner production processes. Some of the more popular uses of collaborative robots are for inspections, sorting, and packaging in collaboration with workers. Robot manufacturers, like Yaskawa, recently announced a new "Smart Series" cobot line that reduces research processes by providing investors with all the necessary accessories to guarantee fast setup. More industries are integrating cobots, manufacturing and non-manufacturing, as a quick and affordable turnkey solution.

Mobile Robot Applications

The most significant influence of supply chain industries has been the tremendous growth of e-commerce sales following the global pandemic. Robots are used across all parts of the supply chain, but as e-commerce fulfillment increases, mobile robots in warehouses and distribution centers are inevitable. These robots autonomously change routes without human intervention and can operate collaboratively with workers to perform tasks. In addition, technologies like voice-picking provide companies with faster production time and flexibility in workflow. Voice Picking is a software that gives AMRs the capability to operate by giving verbal commands to the mobile robots. Companies using voice-picking AMRs report having immediate gain in productivity and savings with annual labor costs. The notoriety of mobile robots is exploding because of the flexible task handling they provide that other automation solutions and manual labor do not.

What Are the Capabilities of Automation in the Future?

The recent surge in automation was a response to overcome supply shortages caused by Covid-19. The drastic increase in demand within such a short time frame made it difficult for suppliers to keep up without automating. Last year, in 2020, A3 reported robot orders from life science industries increased by a momentous 69% compared to the previous year. Market analysts project that life science industries, among other sectors, will continue to see more companies automating within the next few years. With more investments being put into automation, advancements will come at a quicker rate. Automation is now blending traditional industrial robots with IoT (Internet of Things) technology to create lean production processes further and collect data for the systems to improve efficiency autonomously. Converging these two technologies into one system physically automates and digitally records data that can be used for machine learning and improving processes. Future automation will feature more of this blended technology, offering consumers cutting-edge automation to further enhance their supply chain process.

(873 words)

https://www.robots.com/articles/most-popular-industrial-robotic-applications-for-2021-and-projections

New Words and Expressions

accessory [əkˈsesəri] n.	配件；附件
cobot [kəubət] n.	协作式机器人，人机合作机器人
converge [kənˈvəːdʒ] v.	聚集，汇集
cutting-edge a.	领先的，尖端的
intervention [ˌintəˈvenʃn] n.	干涉，干预
notoriety [ˌnəutəˈraiəti] n.	恶名，坏名声
overhaul [ˈəuvəhɔːl] n. & v.	全面改革；彻底检修
pandemic [pænˈdemik] n.	流行病，瘟疫
revenue [ˈrevənjuː] n.	财政收入，税收收入，收益
sanitation [ˌsæniˈteiʃn] n.	环境卫生；卫生设备
syringe [siˈrindʒ] n.	喷射器；注射器；吸管
warehouse [ˈweəhaus] n.	仓库，货仓
workflow n.	工作流程
collaborative [kəˈlæbərətiv] robot	协作机器人
lean production	精益生产
spectral [ˈspektrəl] imaging	光谱成像
supply chain	供应链
turnkey solution	整体解决方案，交钥匙解决方案

Notes

1. AMR (Autonomous Mobile Robot)   自主移动机器人
2. Association for Advancing Automation (A3)   推进自动化协会
3. IoT (Internet of Things)   物联网
4. Yaskawa   日本安川电气

Text Understanding

I. Choose the best answer according to the passage.

1. In the following four choices, _____ is not the challenge brought by the Covid-19 pandemic.

  A. supply shortages   B. demand increases

C. new safety precautions   D. a major employment gap
2. Spectral imaging detecting color spectrums that human workers cannot perceive is a typical example of _____.
   A. sorting application   B. quality inspection application
   C. collaborative application   D. mobile robot application
3. More popular uses of collaborative robots are for _____ in collaboration with workers.
   A. inspections   B. sorting   C. packaging   D. all of them
4. Automation for _____ offers a long-term solution for labor shortages, social distancing requirements, and supply chain overhaul for businesses.
   A. sorting applications   B. quality inspection applications
   C. collaborative applications   D. mobile robot applications
5. As e-commerce fulfillment increases, _____ in warehouses and distribution centers are inevitable.
   A. sorting robots   B. quality inspection robots
   C. collaborative robots   D. mobile robots

II. Fill in the blanks with proper words according to the passage.

1. Companies redesign business models and supply chain management to develop _____ with higher output to overcome the challenges brought by Covid-19.
2. _____ is used to eliminate defects and ensure customers receive products that meet their expectations.
3. Many industries are integrating cobots, manufacturing and non-manufacturing, as a quick and affordable _____.
4. The most significant influence of supply chain industries has been the tremendous growth of _____ following the global pandemic.
5. Voice Picking is a software that gives AMRs the capability to operate by _____ to the mobile robots.

III. Give brief answers to the following questions.

1. What is the major change in business sectors brought by Covid-19?

2. What are the examples of the sanitation tasks carried out by cleanroom robots in manufacturing?

3. What is the result of using voice-picking AMRs?

Vocabulary Building

IV. Fill in the table below by giving the corresponding translation.

English	Chinese
	质量检验
	精益生产
collaborative robot	
workflow	
	供应链
	社交距离
spectral imaging	
cutting-edge technology	

Translation Practice

V. Translate the following sentences from the passage into Chinese.

1. With many products facing supply shortages, industries such as manufacturing, life science, food processing, retail, and delivery/fulfillment have looked to automation as the solution. This has led to the automation of new applications and advancements to common applications.

2. Vision technology gives an extra layer of security in ensuring all products leaving production are of the highest quality to consumers.

3. Companies are implementing collaborative technology in strategic ways to overcome employment gaps, social distancing restrictions, and the immediate need for leaner production processes.

4. Robots are used across all parts of the supply chain, but as e-commerce fulfillment increases, mobile robots in warehouses and distribution centers are inevitable. These

robots autonomously change routes without human intervention and can operate collaboratively with workers to perform tasks.

_____
_____
_____

In-Class Reading

Section B
What Is Nanotechnology?

Defining Nanotechnology

The term "nanotechnology" has been getting a lot of attention in the media in the past few years. News stories have heralded nanotechnology as the next scientific revolution—with promises of faster computers, cures for cancer, and solutions to the energy crisis, to name a few. But what exactly is "nanotechnology"? And, can it fulfill all these promises?

The formal definition of nanotechnology from the National Nanotechnology Initiative (NNI) is as follows:

Nanotechnology is the understanding and control of matter at dimensions between approximately 1 and 100 nanometers, where unique phenomena enable novel applications. Encompassing nanoscale science, engineering, and technology, nanotechnology involves imaging, measuring, modeling, and manipulating matter at this small scale.

The NNI definition can be distilled to three basic concepts:

1. Nanotechnology is very, very small.

When something is on the nanoscale, it measures between 1-100 nanometers (nm) in at least one of its dimensions. When things are this small, they are much too small to be seen with our eyes, or even with a typical light microscope. Scientists have had to develop special tools, like scanning probe microscopes to see materials that are on the nanometer size scale.

Some materials have always been on the nanoscale—like water molecules or silicon atoms. However, recently scientists have been able to use new tools and processes to synthesize and manipulate materials common at the macroscale to this size, like particles of Titanium Dioxide (TiO_2).

2. At the nanoscale, materials may behave in different and unexpected ways.

At the nanoscale, many common materials exhibit unusual properties, such as remarkably lower resistance to electricity, lower melting points, or faster chemical reactions.

For example, at the macroscale, gold (Au) is shiny and yellow. However, when the gold particles are 25 nm in size, they appear red. The smaller particles interact differently with light, so the gold particles appear a different color. Depending on the size and shape of the particles, gold can appear red, yellow, or blue.

Another example of nanoparticles appearing different than the corresponding macroscale material is found in sun screens. TiO_2 has been used in sun screens and sun blocks for a long time. It is one of the ingredients that make the creams appear white in color. Manufacturers are now using nanoparticles to create creams and gels that are clear—because nanoparticles of TiO_2 appear transparent.

Other properties can change when materials are on the nanoscale, too. For example, aluminum (Al) is the shiny pliable metal used to make soda cans. At the nanoscale, aluminum particles are extremely reactive and will explode. Nanoparticles are more reactive because they have more surface area than macroscale particles.

3. Researchers want to harness these different and unexpected behaviors to make new technologies.

By harnessing these new behaviors, researchers in many different disciplines hope to create many new things ranging from everyday products such as antimicrobial socks and lighter tennis rackets to state of the art solar cells, faster and smaller computers, or medical treatments that selectively treat the disease. Many scientists and engineers think that the possibilities are endless.

Is the IPod Nano an Example of Nanotechnology?

Nanotechnology-enhanced products are already finding their way to the market. However, not every product labeled "nano" really utilizes nanotechnology. Some manufacturers are using the prefix "nano-" to communicate the relative small size of their specific products to potential buyers. In addition, they hope to capitalize the buzz and excitement that has surrounded nanotechnology. For example, Indian car makers recently introduced the Tata Nano (Fig. 9-2).

This car is definitely not on the nanoscale, but the designers chose to name it "Nano" to communicate to potential buyers its use of "high technology" and its small size. This isn't the only example on the market today. A more common example is the iPod

Nano (Fig. 9-3). Again, this music player is not "nano-sized", but it is much smaller than the original iPod. However, it does utilize nanotechnology in the chips and circuitry that make it work—the same way your laptop does.

Fig. 9-2 Tata Nano

Fig. 9-3 iPod Nano

There are some products out there that do use nanotechnology, even if it is not contained in the name. One of the most common instances is the use of silver nanoparticles in consumer products. Silver is inherently antimicrobial and has been used to control bacteria since ancient times. By incorporating nanoscale silver into textiles, plastics, and household appliances, manufacturers can make materials that use a small amount of silver to kill bacteria without affecting other properties of the products. However, there is concern that increased use of silver in this manner may pose an environmental risk or lead to the development of silver-antibiotic-resistant bacteria.

The Project on Emerging Nanotechnologies has created an inventory of consumer products that they believe actually utilize nanotechnology in some way. However, due to the proprietary nature of these products, a definitive decision on exactly how the product is utilizing nanotechnology can often not be made.

Can Nanotechnology Live up to the Hype?

Nanotechnology has been dubbed "the next big thing". It is being heralded as the key to new cancer treatments, energy independence, improved electronics, and bringing clean water to third world countries. With such a diverse range of possible applications, nanotechnology has the capacity to change the world we live in, in the same way that computers have changed society over the last 30 years. But to accomplish these feats, scientists and engineers that work on nanoscale materials still have to better understand how nanoscale materials behave and how to synthesize them reliably.

Some applications are closer to market than others. For instance, researchers, like Prof. Naomi Halas and Jennifer West at Rice University have collaborated to develop a cancer treatment using gold nano shells. The treatment is still under development, but may find its way to clinical trials in the near future.

(978 words)

https://education.mrsec.wisc.edu/what-is-nanotechnology-defining-nanotechnology/

New Words and Expressions

antimicrobial [ˌæntimaiˈkrəubiəl] a. & n.	抗菌的 & 抗菌剂,杀菌剂
capitalize v.	利用;积累资本;首字母大写
circuitry [ˈsəːkitri] n.	电路系统,电路装置
distill [diˈstil] v.	提取;蒸馏
dubbed [dʌbd] a.	被称为的;译制的
gel [dʒel] n.	凝胶
herald [ˈherəld] v.	预示……的来临;公开称赞
hype [haip] n.	大肆宣传
inventory n.	清单;存货
nanometer [ˈnænəumiːtə] n. (abbr. nm)	纳米(即十亿分之一米)
pliable [ˈplaiəbl] a.	易弯曲的,可塑的,柔韧的
proprietary [prəˈpraiətri] a.	专利的,专营的;所有权的
clinical trial	临床试验

light microscope　　　　　　　　　　光学显微镜
sun block　　　　　　　　　　　　　防晒霜
sun screen　　　　　　　　　　　　　隔离霜,防晒液

Notes

1. National Nanotechnology Initiative (NNI)　　国家纳米技术倡议
2. Titanium Dioxide (TiO_2)　　　　　　　　二氧化钛(物理性防晒成分)

Text Understanding

Ⅰ. Fill in the blanks with proper words according to the passage.

1. Since nanoscale materials are extremely small, scientists have to use _____ to see the materials.
2. At the nanoscale, many common materials exhibit some _____.
3. Manufacturers make materials that use a small amount of nanoscale silver to _____ without affecting other properties of the products.
4. Nanotechnology is believed to change the world we live in, in the same way that _____ have changed society over the last 30 years.
5. Prof. Naomi Halas and Jennifer West at Rice University have collaborated to develop a tentative cancer treatment by using _____.

Ⅱ. Give brief answers to the following questions.

1. What's the formal definition of nanotechnology from the National Nanotechnology Initiative?

2. What are the three basic concepts of the NNI definition?

3. Why did Indian car makers name their car the Tata Nano since it has nothing to do with nanotechnology?

Vocabulary Building

Ⅲ. Fill in the table below by giving the corresponding translation.

English	Chinese
	防晒霜
	光学显微镜
circuitry	
gel	
	纳米
	临床试验
antimicrobial	
inventory	

Translation Practice

Ⅳ. Translate the following sentences from the passage into Chinese.

1. News stories have heralded nanotechnology as the next scientific revolution—with promises of faster computers, cures for cancer, and solutions to the energy crisis, to name a few.

2. By harnessing these new behaviors, researchers in many different disciplines hope to create many new things ranging from everyday products such as antimicrobial socks and lighter tennis rackets to state of the art solar cells, faster and smaller computers, or medical treatments that selectively treat the disease.

3. Nanotechnology has been dubbed "the next big thing". It is being heralded as the key to new cancer treatments, energy independence, improved electronics, and bringing clean water to third world countries.

Section C
3D Printing: History, Processes and Future

Today's 3D printers had their start in the rapid prototyping (RP) technologies of the 1980s and found their use in the industrial market. As patents expire, 3D printing technology is becoming more available to consumers.

Beginning of an Industry

At the time of their invention in the 1980s, 3D printers focused on creating scale models using computer-aided design (CAD). The first patent for a 3D printer was issued to Charles Hull for "Apparatus for Production of Three-dimensional Objects by Stereolithography" in 1986. Hull went on to start the first 3D printing company: 3D Systems Corporation.

With this background in industrial manufacturing, most 3D printing technology was protected by patents held by companies in the industrial market. As patents expired, 3D printers geared for the consumer began to appear.

Dr. Bryony Core, a Technology Analyst at IDTechEx, explains: " 3D Printing technologies have existed since the 1980s; however, it was not until the expiration of Stratasys' patent in 2009 protecting their ownership of the fused deposition modeling (FDM) process that affordable consumer thermoplastic extrusion printers proliferated. The expiration of subsequent patents covering alternative additive manufacturing processes has further fuelled this growth."

The 3D Printing Process

3D printers build objects using a process known as additive manufacturing. Material is put down in layers; each layer adds to the previous layer and in turn becomes a base for the next layer. Most 3D printers in the consumer market use thermoplastic inks in the printing process. These polymers become soft and pliable within a temperature range and then re-solidify when allowed to cool.

Referring to Fig. 9-4, the print bed is where the object is printed. It's usually covered with an adhesive material, and with some inks it also needs to be heated in order to minimize distortion in the printed object. The extruder temperature may be set manually, depending on the model; the nozzle position is controlled by the microcontroller, which is directed by commands in the print file.

Some printers incorporate USB ports to read files from USB drives; other printers interface to external computers, which may be running 3D print monitor and control applications. The microcontroller positions the nozzle at the X, Y, and Z coordinates needed and a specific amount of ink is set. High-precision 3D printers produce minimal wasted material.

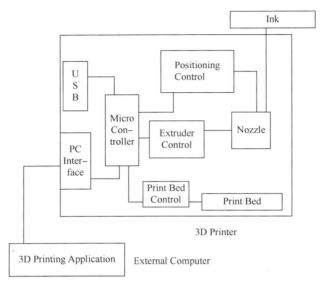

Fig. 9-4 A functional diagram of a 3D printer

3D Printer Inks

Fig. 9-5 shows a variety of filaments, the most common form of ink used by 3D printers. Polylactic Acid (PLA) and Acrylonitrile Butadiene Styrene (ABS) filaments are usually used in consumer 3D printers. Filaments are categorized by diameter, extruder temperature required, print bed temperature needed, and what the print bed should be coated or covered with for best adhesion.

Fig. 9-5 PLA filaments

Print File Formats

Directions for positioning the nozzle and controlling the ink(s) are received from a print file. There are many file formats in use. While CAD programs have been the traditional way to produce files, today newer design tools, as well as scanners and cameras, are used.

These tools can generate and capture data not needed for a 3D print of the object, and their output may be in a proprietary format. As shown in Fig. 9-6, a file may need to be translated to an accepted format for 3D printing. This may involve taking out extra data, modifying the file format, and slicing (layering) the data. Software programs such as blender, an open source 3D creation suite, will accept and translate many file formats.

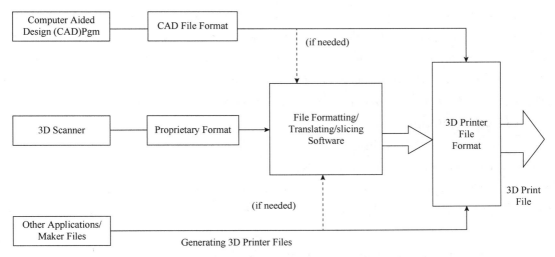

Fig. 9-6 Generating 3D print files

Table 9-1 lists a few of the 3D print file formats in use. The *.STL format was created to support Chuck Hull's original printer. It supports a layering process for one extruder. As printers developed, other formats, both open-source and proprietary, were developed. In 2015, the 3D Manufacturing Format (*.3MF) was announced. Developed and supported by the 3MF Consortium, the aim is to standardize 3D print files.

Some 3D Print File Formats — Table 9-1

*.STL	Stereolithography; 3D single color (one extruder- 1980s)
*.WRL	Vermal; multi-color (supports more than one extruder)
*.OBJ	Open source 3D geometric format
*.X3G	MakerBot format
*.PLY	Scanner output
*.3MF	3D manufacturing format - 3MF Consortium specification (introduced in 2015, and still in use)

Today's Limits

Within the context of industrial manufacturing, 3D printers face limits in what can be manufactured as well as uncertainty within the legal framework surrounding print files and printed objects. Unlike traditional manufacturing technologies that turn out hundreds of components per hour, 3D printing is a slow process.

Beyond the printer technology, providing the file for a 3D-printed object requires expertise in both modeling and software. After printing is completed, a certain amount of handling is required to remove the printed object and finish it. Currently, materials are limited and some printing technologies may end up not being a good fit for all markets.

Questions surround liability of parts when there's a failure, as well as how to handle the intellectual property involved. If a part is printed in a truck outside your house, as an Amazon patent hints at, is the liability the same as a part shipped from a vendor's manufacturing facility? As these issues get sorted out, there is an exciting future for 3D printing, well beyond what was imagined when Chuck Hull filed his patent.

An Exciting Future

Industries using 3D printing have grown with improvements in capturing 3D data, control accuracy, and available materials. The 3D Printing Conference & Expo, held in New York City in March 2017, offered sessions in bioprinting, automotive, aerospace, consumer products, medical, oil and gas, hobbyist/maker, dental, education, and jewelry.

As patents expire, new inks and print technologies are being offered. An exciting development for electrical engineers is the availability of conductive inks, which makes it possible to print objects that can conveniently incorporate electronic components such as sensors or LEDs. Magnetic filaments are also available, so prints will stick to magnets.

With composite materials available in filament, liquid, metal, and powder forms, 3D printed objects are moving beyond prototyping to functional and even manufactured areas. Printers with more than 3 axes are printing objects suitable for high-end components. Other techniques allow printing outside the printer frame.

(1038 words)

https://www.allaboutcircuits.com/news/introduction-to-3d-printing-history-processes-and-market-growth/

New Words and Expressions

axes ['æksiz] *n.* 坐标轴(axis 的复数)
bioprinting *n.* 生物打印

composite [ˈkɔmpəzit] n.	复合材料
consortium [kənˈsɔːtiəm] n.	联盟；联营企业
coordinate [kəuˈɔːdineit] n.	坐标
extruder [eksˈtruːdə] n.	（塑料）挤出机
filament n.	长纤丝；灯丝
gear v.	适合
high-end a.	高端的，高档的
liability [ˌlaiəˈbiləti] n.	责任，义务
proliferate [prəˈlifəreit] v.	激增，增殖
stereolithography [ˈsteriəuˌliˈθɔgrəfi] n.	立体光刻，立体平版印刷
additive manufacturing	增材制造，叠加制造，增量制造
intellectual property	知识产权
open source	开放源码，开源

Notes

1. Acrylonitrile Butadiene Styrene (ABS)	丙烯腈—丁二烯—苯乙烯共聚物
2. computer-aided design (CAD)	计算机辅助设计
3. fused deposition modeling (FDM)	熔融沉积成型
4. Polylactic Acid (PLA)	聚乳酸
5. Stratasys	美国斯特拉塔西斯公司（3D 打印和增材制造解决方案的全球领先提供商）

Exercises

📖 Text Understanding

I. Decide whether the following statements are true (T) or false (F) according to the passage.

1. The first patent for a 3D printer was issued to Charles Hull in 1986, and later he started the first 3D printing company.
2. 3D printing technology was not affordable to consumers until the expiration of the related patents in the 21st century.
3. Most 3D printers in the consumer market use thermosetting inks in the printing process.

4. The most common forms of inks used by 3D printers are PLA and ABS filaments.

5. CAD programs, scanners and cameras are all new ways to produce print files.

6. Since 3D printers have different file formats, they have to be translated to an accepted format before 3D printing.

II. Give brief answers to the following questions.

1. What is the printing process of 3D printers? Explain it in detail.
___

2. How do 3D printers read files to perform tasks?
___

3. What are the limits 3D printers will face today?
___

Vocabulary Building

III. Match the items listed in the following two columns.

1. coordinate　　　　　　　a. a coming to an end of a contract period

2. expiration　　　　　　　b. intangible property that is the result of creativity, such as patents, trademarks, or copyrights

3. composite　　　　　　　c. a document granting an inventor sole rights to an invention

4. patent　　　　　　　　　d. an early sample, model, or release of a product built to test a concept or process or to act as a thing to be replicated or learned from

5. intellectual property　　　e. a thing made up of different parts or materials

6. prototype　　　　　　　f. a number that identifies a position relative to an axis

Translation Practice

IV. Translate the following paragraph into English.

　　3D打印即快速成型技术的一种,又称增材制造,是一种以数字模型文件为基础,运用粉末状金属或塑料等可黏合材料,通过逐层打印的方式来构造物体的技术。3D打印通常是采用数字技术打印机来实现的。常在模具制造、工业设计等领域被用于制造模型,后逐渐用于一些产品的直接制造,已经有使用这种技术打印而成的零部件。该技术在珠宝、鞋类、工业设计、建筑、工程和施工、汽车、航空航天、牙科和医疗产业、教育、土木工程、枪支以及其他领域都有所应用。

Translation Skill IX

句法的处理(3)—— 从句的翻译

科技英语翻译时,英语从句由于其句长及内在逻辑增加了翻译的难度。如何精准翻译出科技英语中的各种从句直接关系到整个科技英语语篇的翻译质量。在下文中,我们将重点介绍英语中出现频率较高的定语从句和状语从句的翻译技巧。

一、定语从句的翻译

英语的定语从句是指被关系副词或关系代词所引导的从句,在句子中充当定语的成分,通常修饰、限制某一个代词或名词,被修饰的词叫作先行词,其后所接的即是定语从句。虽然从句法角度来说,定语从句属于次要部分,但从语言角度来说,它具有重要意义。在翻译定语从句时,不提倡一概将定语从句翻译成汉语的修饰语,需要对句子的结构、意义进行具体分析,适当调整、转换,综合运用定语从句的翻译技巧,这样才能译出通顺、流畅,符合汉语表达习惯的译文。总的来说,定语从句的翻译技巧可以概括为以下几个方面:

1. 采用前置法

当英语定语从句比较简短时,可将其译成带"的"的定语短语,放在被修饰词之前,从而将英语的复合句译成汉语的简单句。

A coil of wire that moves in a magnetic field will have all e.m.f. induced in it.

译文:在磁场中运动的线圈会产生感应电动势。

The pumps which are used on board ship can be divided into two main groups.

译文:在船上使用的泵可分为两类。

2. 译成并列句

定语从句比较长,意义上独立性较强时,可把定语从句拆译成并列的独立句,这时可重复先行词或用"这""那""其""该"等作为独立句的主语。

Centrifugal pumps consist of an impeller, which rotates at high speed, inside a casting.

译文:离心泵有泵叶轮,叶轮以高速在泵壳内转动。

Some metals are better conductors of electricity than others, which means that the former have atoms that contain more free electrons than the latter.

译文:某些金属比另一些金属有更好的导电性,这表明前者所拥有的原子,其所含自由电子数比后者要多。

3. 译成状语从句

英语定语从句与主句之间有着较为复杂的关系,有时只从语法角度分析从句与先行词之间的关系不利于理解和翻译,这是因为有的定语从句修饰作用极弱,其成分和状语作用类似,常常包括原因、结果、目的、让步、条件、时间等意思。因此,要想译好此类定语从句,必须弄清主句与从句间的逻辑关系,将带有状语功能的定语从句转化为恰当的状语进行翻译。

A.译成原因状语

Einstein, who worked out the photoelectric effect, won the Nobel Prize in 1921.

译文:爱因斯坦因为提出了光电效应理论而获得了1921年诺贝尔奖。

B.译成结果状语

Copper, which is used widely for carrying electricity, offers very little resistance.

译文:铜的电阻小,因此被广泛用来输电。

C.译成目的状语

The bulb is sometimes filled with an inert gas which permits operation at a higher temperature.

译文:有时往灯泡内充入惰性气体,是为了能在更高的温度下工作。

D.译成让步状语

Friction, which is often considered as an obstacle, is of much help to the vehicles moving forward.

译文:摩擦虽然被看作是一种障碍,但对车辆行进却大有裨益。

E.译成条件状语

For any machine whose input force and output force are known, its mechanical advantages can be calculated.

译文:对于任何机械,只要知晓其输入力和输出力,就能求出其机械效益。

F.译成时间状语

Very loud sounds produced by huge planes, which fly low over the land, can cause damage to house.

译文:巨型飞机低空飞行时产生的巨大轰鸣声,可能对房屋造成破坏。

4.其他翻译方法

英语中的定语从句还有一些其他的翻译处理方法,比如和主句合译成简单句、加括号译出等。

There are some engine rooms on board seagoing ships where all the devices are automatically-controlled.

译文:远洋轮船上有些机舱设备是自动控制的。

Some materials, such as cotton, which is often used as insulation, are liable to absorb moisture, and this will adversely affect their insulating properties.

译文:有些物质(如常用作绝缘体的棉花)容易吸水,这对其绝缘性能也有不利的影响。

二、状语从句的翻译

英语的状语从句是指用来修饰主句或主句谓语的英语句子。状语从句一般可以分为九大类,分别表示时间、地点、原因、目的、结果、条件、让步、比较和方式。引导状语从句的连词大多可以在汉语中找到相对应的表达。翻译状语从句时,可以根据不同类型,采取顺序译法、逆序译法和变序译法,但对于一些比较特别的句子,还需综合运用翻译技巧,将其译为不同类的状语从句,或直接将其与主句合译成简单句。下面,我们将对以上翻译技巧分别举例进行阐述:

1. 顺序译法

在汉语表达中,偏正复句中的偏句一般在前,因此当英语原文中状语从句位于主句之前时,所翻译的句子可以直接按顺序译出,即顺序译法。

When solar energy penetrates the land surface, it is converted into heat, most of which radiates upward quickly.

译文:当太阳能照射陆地表面时,它就转化为热能,其中大部分迅速向上辐射。

Even though automatic machines can do many things that man does, they cannot replace man.

译文:虽然自动化的机器能做人所做的许多事,但是它们不能代替人。

2. 逆序译法

A. 英语中的状语从句通常都位于主句之后,但在翻译时,由于我们受汉语语言惯用表达方式的影响,会将其译在主句之前,即采用逆序法译出。

Some sulfur dioxide is liberated when coal, heavy oil, and gas burn, because they all contain sulfur compounds.

译文:因为煤、重油和煤气都含有硫化物,所以它们燃烧时会放出一些氧化硫。

The selectivity of the electrode increases as the concentration of organic salts decreases.

译文:当有机盐的浓度降低时,电极的选择性就增大。

B. 有时,当目的状语从句、条件状语从句、原因状语从句和结果状语从句位于主句之后时,也可以按照原文的顺序将其译出,以保持上下文的一致。

Engineers today plan swiftly and effectively because they know how to measure and calculate.

译文:现在,工程师做设计速度快、效率高,因为他们知道如何测量和计算。

D-mode is applied most in common drive, so it needs to highlight best fuel consumption.

译文:驱动器模式是最常见的驱动方式,所以它需要强调最佳的燃油消耗。

3．变序译法

有时为了符合汉语的表达习惯,可将英语中的状语从句插入汉语的主谓之间。

The earth turns round its axis <u>as it travels around the sun</u>.

译文:地球<u>一面绕太阳公转</u>,一面自转。

Some materials are called good conductors <u>because electricity goes through them well</u>.

译文:有些材料<u>因导电性能好</u>被称为良导体。

4．转换成其他从句

有些状语从句从语法再到语义逻辑,从不同视角理解,就会呈现出不同于引导词的其他状语从句。

A．转换成条件状语从句

<u>Where water resources are plentiful</u>, hydroelectric power stations are being built in large numbers.

译文:<u>只要是水源充足的地方</u>,就可以修建大批水电站。

B．转换成原因状语从句

<u>When winds blow sand particles against a large rock for a long time</u>, the softer layers of the rock are slowly worn away.

译文:<u>由于风把砂砾吹起来,碰撞大岩石</u>,因而较松散的岩石层便会慢慢磨损。

5．和主句合译成简单句

状语从句有时也可以一言以蔽之,缩减翻译成为句子中的一部分,和主句合译成为简单句。

The induced e.m.f. is in such a direction <u>that it opposes the change of current</u>.

译文:感应电动势的方向<u>与电流的变化方向相反</u>。

Electricity flows through a wire just <u>as water flows through a pipe</u>.

译文:电流通过导线就像<u>水流过水管一样</u>。

上文介绍了定语从句和状语从句的翻译技巧,在实际翻译中,我们还可能会遇到主语从句、宾语从句、同位语从句等各种不同类型的从句。需要指出的是,没有哪一种翻译技巧是万能的,我们只有在充分阅读理解文本的基础上,灵活运用各种翻译技巧,通过不断实践,逐步提高自身的翻译水平,尽可能地为读者呈现出翻译精准且接受度较高的译文。

PART IV
Prospect

Unit 10

Prospect of Electrical and Mechanical Engineering

Before-Class Reading

Answer the following question before reading the passage:

1. Try to describe at least one future development in electrical or mechanical engineering in your own words.

Section A
Electrical and Electronic Engineering in the Future

So much of the developing of science and technology depends on the variables of economic, political, and social developments that precise predications about future trends in the field of electrical and electronic engineering cannot be made. But it is possible to discern certain technological trends which can reasonably be expected to occur.

Experts on the future are divided into pessimists and optimists. Pessimists forecast doom. They point to increasing pollution of the atmosphere, water, and land, the depletion of raw materials, the exhaustion of some now-common energy sources, and the geometric expansion of population.

Optimists foresee unlimited energy through harnessing the power of the atom, the discovery of new and unlimited food sources, and the dawn of an age in which human drudgery is replaced by technological advances. The truth is probably somewhere in between.

A major scientific advance such as the development of a comprehensive theory and knowledge of elementary particles, the basic components of all matter and energy, could profoundly change the way we live tomorrow. A theory based on knowledge of such strange objects as pulsars(Fig. 10-1), stars that emit radio waves in uniform pulses, and quasars, strong radio sources and unexplained sources of enormous energy, could alter life in ways we cannot yet forecast.

Fig. 10-1　A constellation of pulsars

Electrical and electronic scientists and engineers are engaged in examining and developing these areas. They are not restricted in their exploration to our present knowledge of space or time. With an instrument known as an electron accelerator, they probe the mysteries of the atomic nucleus, and with the radio telescope they study signals from remote regions of outer space. With computers they can store information indefinitely and with electronic circuits they can get information in a thousand-billionth of a second.

A major success such as the harnessing of thermonuclear energy produced through nuclear fusion would radically affect the development of all branches of engineering. The world would move from a state of energy scarcity to an era of inexhaustible energy resources. Given the proper economic and political circumstances, this would cause tremendous growth in science and technology.

However, if this major breakthrough does not occur, the enormous need for new energy resources will continue to grow. With the ever-increasing use of fossil fuels, the effort of much of the technological community is already directed toward the discovery of new sources of non-fossil fuels. Solar, geothermal, tidal, and wind sources of energy are gradually becoming more economically possible. A technological breakthrough in any one of these fields would provide research work for tens of thousands of electrical engineers. New and improved types of cells, batteries, generators, converters, power plants, and transmission lines would have to be designed, tested, evaluated, and built in order to properly use the new source of energy.

In any case, the future of electrical and electronic engineering does not depend solely on the development of new scientific theories or the discovery of new energy sources. These engineers will be engaged in diverse technological pursuits such as the following:

Electrical and electronic engineers will be intimately involved in the development of the completely automated industrial factory. It will become possible, with the aid of electronic computers, to produce goods by teams of machines that transfer materials from one to another. In such a factory a product could be manufactured, tested, labeled, packaged, and shipped without being touched by human hands or directed by human intellect.

In the field of transportation, electrical engineers are currently engaged in developing the electric automobile, train, bus, and ship. They are designing new inertial guidance systems which would guide rockets and interplanetary spaceships by using devices which detect changes in speed and direction and make necessary adjustments automatically.

Fueling aircraft and spacecraft by laser beam is another possibility that will transform future travel. As light energy can be converted into other forms of energy, so could the laser beam be converted to aircraft fuel. Such a breakthrough would greatly reduce the weight of aircraft and thereby increase the probability of hypersonic travel—travel at speeds five or more times greater than the speed of sound. Planes could travel at 4,000 to 5,000 miles an hour and at altitudes of 150,000 feet.

Society will become more and more computerized, and the electronic engineer will be called upon to design and build ever-smaller computers capable of doing more varied and more complicated tasks. At some time in the future, fully automatic automobiles and homes will be built and directed by computers. Computers that "think", that learn from errors and never make the same mistakes twice, that are able to repair themselves and reproduce themselves, may be the reality of tomorrow.

Cybernetics, the science of automatic controls, could eventually produce a race of robots—machines in human shapes that perform human tasks with what parallels human intelligence. Only human sensitivity, emotion, and sexuality will be missing. The necessary scientific knowledge for building these labor-replacing devices is available to engineers today: computer technology, microcircuit technology, control theory, and information theory—a mathematical analysis of the efficiency with which computers, telecommunication channels, and other information-processing systems are employed. The electronic engineer need only translate today's knowledge into tomorrow's machinery.

The exciting field of biomedical engineering offers enormous possibilities. More and more electronic instruments to extend, repair, and improve upon physical life are currently being developed. Lasers are already used to join living tissues such as detached eye retinas; their uses in surgery too intricate and delicate for the knife will become commonplace. Computers will be developed to diagnose and treat disease. Electronic engineers will devise more usable and varied organs and organ replacements. There is, theoretically, no limit to the uses of electronics in medicine.

Not only will the new developments call for electronic engineers, they will develop new electronic products for people to buy as well. Telephones with picture screens on which the connected parties can see each other and three-dimensional television which would completely envelop the viewer could become ordinary household items.

Research and development, or R&D—investigation and experimentation by scientists, engineers, and technicians—is not confined to sciences such as physics or radio astronomy. Countless engineers will continue to design and improve upon existing vacuum tubes, switches, and electromechanical devices. Improvements will be made in antennae, arrangements of wires and rods which fan out to receive electromagnetic waves; filters, which block out selected waves or current; transducers, which convert one form of energy to another; and relays, which electrically cause switches in a circuit to open and close. These are the basic components of the electronics industry and a vital segment of the industries that maintain our economy.

These exciting possibilities indicate a bright future for electrical and electronic engineers. They will play a central role in formulating, shaping, and bringing into the immediate and distant future.

(1137 words)

From *English for Machinery*

by Bu Yukun

New Words and Expressions

antennae [æn'teni:] n.　　　　　　　　天线;触角
cybernetics [ˌsaibə'netiks] n.　　　　　控制论
depletion [di'pli:ʃən] v.　　　　　　　耗尽,干枯,枯竭
discern [di'sə:n] v.　　　　　　　　　分辨,觉察,识别
drudgery ['drʌdʒəri] n.　　　　　　　苦差事;单调沉闷的工作
envelop [in'veləp] v.　　　　　　　　包围;遮盖
geometric [ˌdʒi:ə'metrik] a.　　　　　几何学的
hypersonic [ˌhaipə'sɔnik] a.　　　　　特超音速的
inexhaustible [ˌinig'zɔ:stəbl] a.　　　用之不竭的
pulsar ['pʌlsɑ:] n.　　　　　　　　　脉冲星
quasar ['kweizɑ:] n.　　　　　　　　类星体;类星射电源
retina ['retinə] n.　　　　　　　　　视网膜
scarcity ['skeəsəti] n.　　　　　　　　缺乏,不足
transducer [trænz'dju:sə] n.　　　　　换能器,转换器
electron accelerator　　　　　　　　电子加速器

inertial guidance system	惯性导航系统,惯性制导系统
interplanetary spaceship	星际飞船
living tissue ['tiʃuː]	生物组织,活组织
radio telescope	射电望远镜

📖 Text Understanding

Ⅰ. Decide whether the following statements are true (T) or false (F) according to the passage.

1. The author views the future as somewhere in between the predications of the pessimists and the optimists.
2. Cybernetics could eventually produce robots that perform human tasks with what parallels human intelligence and emotion.
3. In biomedical engineering, lasers will be used in surgeries too intricate and delicate for the knife.
4. Three-dimensional television which would completely envelop the viewer has become ordinary household items.
5. Electrical and electronic engineers will play a central role in formulating, shaping, and bringing into the immediate and distant future.

Ⅱ. Give brief answers to the following questions.

1. How do optimists forecast the future?

____

2. What will electrical and electronic scientists and engineers do with instruments like electron accelerators and radio telescopes?

____

3. What are inertial guidance systems capable of?

____

4. What are the advantages of fueling aircraft and spacecraft by laser beam?

____

Vocabulary Building

Ⅲ. Fill in the table below by giving the corresponding translation.

English	Chinese
depletion	
cybernetics	
	电子加速器
	变量
interplanetary spaceship	
radio telescope	
	视网膜
	脉冲星
transducer	
	惯性制导系统

Ⅳ. Fill in the blanks with the words from the passage. The first letter of the word is given.

1. Many pictures have been based on simple g_____ designs.
2. She has an i_____ supply of enthusiasm.
3. Either food s_____ or excessive hunting can threaten a population of animals.
4. At nightfall, the fog began to e_____ the campground.
5. One of the panels had become d_____ from the main structure.
6. The aircraft would reach h_____ speed—five times the speed of sound.

Translation Practice

Ⅴ. Translate the following sentences from the passage into Chinese.

1. Pessimists forecast doom. They point to increasing pollution of the atmosphere, water, and land, the depletion of raw materials, the exhaustion of some now-common energy sources, and the geometric expansion of population.

   _____
   _____
   _____

2. Given the proper economic and political circumstances, this would cause tremendous

growth in science and technology.

3. Lasers are already used to join living tissues such as detached eye retinas; their uses in surgery too intricate and delicate for the knife will become commonplace.

4. Improvements will be made in antennae, arrangements of wires and rods which fan out to receive electromagnetic waves; filters, which block out selected waves or current; transducers, which convert one form of energy to another; and relays, which electrically cause switches in a circuit to open and close. These are the basic components of the electronics industry and a vital segment of the industries that maintain our economy.

In-Class Reading

Section B
The Future of Mechanical Engineering: A Vision and a Mission

The future of mechanical engineering has paved way for engineering milestones such as biomechanics, nano-engineering, robotics, microelectronics, etc. What does the future hold for the professional mechanical engineer? Let's look.

The Future of Mechanical Engineering: A Vision and a Mission

Mechanical Engineering has been around for centuries and will be, for a long time to come, unless there is a miracle in science that allows humans to deny all laws of mechanics and still allows them to build stuff that can be used. As of now, the situation is unfathomable.

From basic objects like wheels to the ever useful screws and inclined planes, from cars to aeroplanes, from paperclips to ships, from bridges to skyscrapers, they all work under the foundations and principles laid out by the laws of mechanics.

We have seen how machines have made our lives easier. Thanks to mechanical engineering, they have increased the efficiency of the machines that we use and also made it easier to make them. We have seen the wonders of mechanical engineering, but what is the future of mechanical engineering?

The future of mechanical engineering is spread across various emerging streams that hold many promises to make the future a better place to live in. Some of the promising streams that are the quintessential applications of the future of mechanical engineering are:

1) Nano-technology—Nano-engineering to be more specific
2) Biomechanics—A promising stream of the future
3) Automobiles and aviation
4) Buildings of the future and urban designing
5) Robotics

... and the list continues, but these are the major areas where the future of mechanical engineering will logically be applied. Now let's take a tour of what they are and see how they are in the future.

Nano-engineering

"Nano" is the word used to measure any object that is measured in scales of 10 to the power of -9. They are microscopic and only electron microscopes are used to see such objects. Creating anything of that microscopic size in itself is a stellar task, let alone applying the principles of mechanical engineering at that level. Nano-engineering is mainly used in the field of research and medicine. It is used to create materials that are not affected by normal factors like weather and corrosion, etc. Materials designed by nano-engineering are lighter and stronger than other materials. The mechanical structure of the materials is changed, thereby giving enhanced properties to the materials. A concept using carbon nanotubes by NASA is said to be able to link the surface of the earth and the satellite, thereby making a direct connection to a satellite. This is the potential of nano-engineering.

Biomechanics

Bionics is the boon of modern medical science that promises to make the lives of millions better. "Bionic" is the term used to refer to the artificial material or object that mimics the action done by a part of the human body. For example, a bionic arm mimics the actions of a human arm; the bionic leg(Fig. 10-2) mimics the human leg. They are created using the principles of biomechanics. Experiments are going on that promise bionic chests and bionic necks. The functions and the operations of the bionics that copy their human counterparts heavily depend on the principles laid by mechanical engineering. Bionics is one area where we can see the wonders of what the future of mechanical engineering has in hold for us and how it is being applied in day-to-day life.

Automobiles and Aviation

Mechanical engineering has helped in creating the fastest cars that are capable of traveling 400+ kph (248 mph) and in the making of the most comfortable vehicles on the planet that are used by millions. Huge aircrafts that enable millions every day to reach from one corner of the globe to the other in a matter of hours all are the result of extensive improvement and implementation of mechanical engineering. The strength of the body and the way automobiles and aircrafts are built are results of extensive mechanical engineering and testing. Advancements in mechanical engineering are applied to automobiles to decrease their carbon footprint and make them more eco-friendly and economical while simultaneously giving more efficiency.

Fig. 10-2 A bionic leg

Buildings of the Future and Urban Design

Huge structures like the Burj-Khalifa, Taipei 101, and many other tall buildings use mechanical engineering for the structure of the building. Taipei 101 uses mass dampers for stability so that there is a uniform weight distribution, which ensures that the buildings don't get unstable easily. The mechanical structure of the buildings are so adjusted that they are resistant to winds of high speed and natural disasters like earthquakes, storms, etc. The mechanical structure of the building is built such that any tremors at the base of the building are gently damped and the vibrations do not cause any serious effects. Thus mechanical engineering plays a pivotal role in the construction of buildings along with various other sciences.

Fig. 10-3 ASIMO

Robots

Robots like ASIMO (Advanced Step Innovative Mobility), as shown in Fig. 10-3, can walk, jog, climb stairs, greet people, and do a lot of other things. Robots like ASIMO are the future. To perform all those actions, the robots need to work like humans and mechanical engineering helps in the functioning of the limbs and other body parts. The same principle of biomechanics is used in this area of science. Nano robots are also in the making that are said to be of immense use in the field of medicine, though many oppose the concept as they are also potential weapons of mass destruction and cannot be stopped easily.

Though the wonders of mechanics are many, a very few have been listed, so that a rough idea of what the future of mechanical engineering has in hold for humanity, is perceived.

(973 words)

https://www.brighthubengineering.com/machine-design/72667-the-future-of-mechanical-engineering-a-vision-and-a-mission/

New Words and Expressions

aviation [ˌeiviˈeiʃn] n.	航空
biomechanics [ˌbaiəuməˈkæniks] n.	生物力学
bionics [baiˈɔniks] n.	仿生学
boon [buːn] n.	非常有用的东西；恩惠，益处
microscopic [ˌmaikrəˈskɔpik] a.	极小的，微小的
paperclip n.	回形针，曲别针
pivotal [ˈpivətl] a.	关键的，核心的
quintessential [ˌkwintiˈsenʃl] a.	典型的，本质的，精髓的
skyscraper n.	摩天大楼
stellar [ˈstelə] a.	出色的，精彩的；恒星的
tremor [ˈtremə] n.	轻微地震，小震
unfathomable [ʌnˈfæðəməbl] a.	难以理解的，高深莫测的
carbon footprint	碳足迹
carbon nanotube	碳纳米管
inclined plane	斜面，倾斜面
mass damper	质量阻尼器
weapons of mass destruction	大规模杀伤性武器

Notes

1. ASIMO (Advanced Step Innovative Mobility) —— 阿西莫(高级步行创新移动机器人)
2. Burj-Khalifa —— 哈利法塔，迪拜塔
3. NASA (National Aeronautics and Space Administration) —— 美国国家航空航天局

📖 Text Understanding

I. **Fill in the blanks with proper words according to the passage.**

1. All the objects work under the foundations and principles laid out by _____.
2. "Nano" is the word used to measure any object that is measured in scales of _____.
3. Advancements in mechanical engineering are applied to automobiles to decrease their _____ and make them more _____ while simultaneously giving more efficiency.
4. Taipei 101 uses _____ for stability so that there is a uniform weight distribution.
5. Nano robots are opposed by many people because they might be _____ and cannot be stopped easily.

II. **Give brief answers to the following questions.**

1. What are the specific applications of the future of mechanical engineering mentioned in the passage?
   _____
2. What does the term "bionic" refer to? Please give at least one related example.
   _____

📖 Vocabulary Building

III. **Fill in the table below by giving the corresponding translation.**

English	Chinese
	仿生学
	斜面
mass damper	
biomechanics	
	碳足迹
	航空
ASIMO	
NASA	

Ⅳ. Fill in the blanks with the words from the passage. The first letter of the word is given.

1. Bionics is the b_____ of modern medical science that promises to make the lives of millions better.
2. He has established himself as a p_____ figure in state politics.
3. The robot was programmed to m_____ a series of human movements.
4. It's a s_____ example of using technology for social good.
5. Indeed, dance is the q_____ gesture language.
6. The t_____ is an aftershock of the March 11th quake.
7. They found fossils of m_____ marine plants which suggest that the region was once open ocean, not solid ice.

✎ Translation Practice

Ⅴ. Translate the following passage into English.

本田公司投入无数科技研究心血的结晶——全球最早具备人类双足行走能力的类人型机器人阿西莫（ASIMO），以憨厚可爱的造型博得许多人的喜爱，众多的类人功能也不断地冲击着人们的想象，似乎科幻电影中的情节正在一步步变成现实。ASIMO 身高 1.3 米，体重 48 公斤。它的行走速度是 0～9km/h。早期的机器人如果直线行走时突然转向，必须先停下来，看起来比较笨拙。而 ASIMO 就灵活得多，它可以实时预测下一个动作并提前改变重心，因此可以行走自如，进行诸如"8"字形行走、下台阶、弯腰等各项"复杂"动作。此外，ASIMO 还可以握手、挥手，甚至可以随着音乐翩翩起舞。

💬 Translation Skill Ⅹ

科技英语语篇的翻译

科技英语语篇是为专门的科技英语读者、专门的科技文本用途而选择和组织信息的过程。科技英语的题材不同于评论性的社论，也不同于抒发情感的诗歌、散文或小说，有其专门的信息语篇类型，旨在揭示客观事实、证明理论假设。科学家和工程师将其在科技工程领域的经验和认知投射在科技领域的信息组织结构中，体现出科技语篇的功能性。科技语篇在信息组织结构方面的特殊功能主要由三方面决定：1）语篇信息内容遵循一定的顺序；2）信息之间有内在的层次关系；3）这些有内在关系的信息有其特殊和固定的表达方式。

科技英语语篇翻译同样需要考虑源语语篇（Source Text，简称 ST）与目标语语篇（Target Text，简称 TT）在意义和功能上的对等。在翻译时，将译文置于语篇框架中，从科技语篇的段落主题、功能及结构等角度反思和分析翻译方法。在翻译中采用语篇翻译方法,有利于更好地再现原文的中心思想和总的基调,使译文中心突出、层次分明、内容衔接连贯。

囿于篇幅,这里以段落为切入点探讨科技英语信息。段落是语篇功能和信息的载体,是科技英语语篇的语义单位,也是分析科技英语语篇的基本单位。从整个科技语篇来看,有核心概念段落和辅助性段落。核心概念段落总括核心信息,辅助性段落进一步展开说明和描述核心信息。而在一个段落里,也有核心概念信息,即核心句,形式上表现为概括性的句子,在随后展开的句子中,核心句中的核心词起着统领性作用。核心句的位置不固定,但大多数时候置于段落句首,有时也置于段落中间或末尾。

下面,我们分别选取机械工程语篇和汽车工程语篇各一例,对英语原文和汉语译文平行文本进行讨论和分析。

机械工程语篇 表1

英语原文 ST	汉语译文 TT
① Standard "engine" lathe, which is the type commonly used by machinists for doing general work, is one of the most important tools in a machine shop, because it is adapted to a great variety of operations, such as turning all sorts of cylindrical and taper parts, boring holes, cutting threads, etc. ② Figure 1 below shows a lathe which, in many respects, represents a typical design, and while some of the parts are arranged differently on other makes, the general construction is practically the same as on the machine illustrated.	① 普通车床,由发动机驱动,是机械加工车间最重要的机床之一。② 普通车床适用于多种不同的加工,如车削各种圆柱形和锥形零件、镗孔、车削螺纹等,因此常被技师们用来加工通用零件。 ③ 图1展示了一台车床。④ 虽然部分零件与其他型号的车床在安置上会有不同,但在许多方面都体现出车床的典型设计,其他车床的总体结构与所展示车床的结构几乎是相同的。

表1中文本的目标读者为机械加工的技工,以及在金工车间实习的学生和学徒。通过对表1中英语原文观察看出,源语文本(ST)呈现为两个段落,但每个段落只有一句话。第一段的核心句位于句首,即车床是金工车间一种最重要的工具,随后多种加工功能逐一引发开来,全部详细阐述都放在第一句话里。第二段的核心句同样位于句首,用图示介绍了车床的结构。我们甚至可以推断,在新的段落中,作者会详细介绍车床的组成部分,如床身、床头箱、床尾箱、刀架、拖板等。

在梳理出核心信息后,还应厘清小句间的逻辑关系,并调整认知和翻译思路。虽然英语原文为两个段落两句话,但汉语不可能以超长句的形式对应翻译,因此需借助前一章所讲的从句拆译的翻译技巧,将第一段中的因果关系、第二段中的定语从句进行拆译处理。再者,语篇翻译时,也要注意源语中个别词义的认知理解和对等。比如,在英语源语中,"standard 'engine' lathe",在汉语译文中变成了"普通车床,由发动机驱动"。在英语源语中,"engine"带有引号,译者为了强调,在汉语译文中将名词活用处理为了动词短语。

汽车工程语篇　　　　　　　　　　　　表 2

英语原文 ST	汉语译文 TT
The modern motor vehicle engine burns a fuel to obtain power. The fuel is usually petrol (gasoline) or diesel, although liquid petroleum gas (LPG) and compressed natural gas (CNG) are sometimes used. Specialist fuels have been developed for racing car engines. Motor vehicle engines are known as internal combustion engines because the energy from the combustion of the fuel, and the resulting pressure from expansion of the heated air and fuel charge, is applied directly to pistons inside closed cylinders in the engine. The term "reciprocating piston engine" describes the movement of the pistons, which go up and down in the cylinders. The pistons are connected by a rod to a crankshaft to give a rotary output. (Fig.3).	现代汽车的发动机通过燃烧燃料获取动力。虽然有时会使用液化石油气和压缩天然气,但汽车燃料通常是汽油或是柴油。研究人员为赛车发动机开发了专门的燃料。汽车发动机也被称为内燃机,这是因为燃油燃烧产生的能量,以及由热空气和燃油混合膨胀所产生的压力,直接作用于发动机封闭气缸内的活塞。术语"往复式活塞发动机"就是描述活塞在气缸中的上下运动的。活塞由连杆连接曲轴,产生旋转输出动力。具体如图 3 所示。

　　表 2 中文本的目标读者为汽车技师和对汽车发动机原理感兴趣的普通驾驶者。源语文本(ST)段落中有两个概念核心——燃料和内燃机工作原理。首先,发动机的驱动力来自燃料,而燃料有不同种类,包括汽油、柴油和专门用途的燃料,不同的车辆使用不同的燃料。其次,内燃机工作原理,即燃料点火,做功于活塞,活塞做往复运动,驱动车轮。从文本组织结构来看,先是演绎燃料概念,随后展开和归纳活塞的运动。

　　译文基于上述两个概念核心,分析各个概念关系。在翻译时,这些概念逻辑关系,环环相扣,自然成篇,精准地翻译出了汽车的燃油种类和内燃机工作原理。

　　以上是基于科技语篇功能,对科技英语段落翻译方法的探讨。在翻译科技英语语篇时,首先需要全面把握语篇的信息功能,厘清语篇层次,明晰语篇中概念的逻辑关系,再综合运用我们前几章所讲述的科技英语词汇、句法的翻译技巧,力争翻译出科学的、符合专业要求的译文。

摘选自高巍、高颖超、陈燕如(2016)
《语篇功能视阈下科技英语段落与翻译研究》

Glossary

A

a function of time	时间函数
abrasion [ə'breiʒn] *n*.	(表层)磨损
accessory [ək'sesəri] *n*.	配件;附件
acid ['æsid] *n*.	酸
actuator ['æktjueitə] *n*.	执行器,执行元件;驱动器
additive manufacturing	增材制造,叠加制造,增量制造
adhesive bonding	粘接,黏合剂
ailment ['eilmənt] *n*.	病痛,小病
air gap	气隙
align [ə'lain] *v*.	校准,排整齐
alkali ['ælkəlai] *n*.	碱
alloy ['æləi] *n*.	合金
alloying agent	合金添加剂
alternate ['ɔːltəneit] *a*.	交替的
alternating ['ɔːltəneitiŋ] current	交流电
altitude ['æltitjuːd] *n*.	海拔,高度
aluminum [ə'luːminəm] *n*.	铝
amber ['æmbə] *n*.	琥珀,琥珀色
ammonium chloride [ə'məuniəm 'klɔːraid]	氯化铵
amorphous [ə'mɔːfəs] *a*.	无定形的,无组织的
amplitude modulation	振幅调制
analog ['ænəˌlɔg] *a*. & *n*.	模拟的 & 模拟;类似物
ancillary [æn'siləri] *a*.	辅助的,附加的
android ['ændrɔid] *n*.	人形机器人
angular misalignment	角度误差;角位移
annealing [ə'niːliŋ] *n*.	退火
anode ['ænəud] *n*.	阳极
antenna [æn'tenə] *n*.	天线
antennae [æn'teniː] *n*.	天线;触角

antics [ˈæntiks] n.	滑稽可笑的举止；荒唐行为
antifriction bearing	减磨轴承，滚动轴承
antimicrobial [ˌæntimaiˈkrəubiəl] a. & n.	抗菌的 & 抗菌剂，杀菌剂
anvil [ˈænvil] n.	铁砧
applied load	外加载荷，外施载荷
armature [ˈɑːmətʃə] n.	电枢（电机的部件）
assemblage [əˈsemblidʒ] n.	装配，集合
assembly [əˈsembli] n.	装配，组装；集合
atom [ˈætəm] n.	原子
atomic fission [ˈfiʃn]	原子裂变
atomic number	原子序数
augment [ɔːgˈment] v.	增加，增大
austenite [ˈɔstəˌnait] n.	奥氏体
automatic breaker	自动断路器
aviation [ˌeiviˈeiʃn] n.	航空
axes [ˈæksiz] n.	坐标轴（axis 的复数）
axial [ˈæksiəl] length	轴向长度，轴长

B

back drive	反向传动
backlash n.	后座，后冲
ball and socket joint	球窝关节
ball bearing	滚珠轴承
bar magnet	条形磁铁，磁棒
bar steel	条钢，棒钢
batch production	批量生产
battery pack	电池组
bauxite [ˈbɔːksait] n.	铝土矿，铝矾土
be proportional to	与……成正比
beam coupling	电子束耦合，光束耦合
bearing n.	轴承
Belleville [ˈbelvil] washer/spring	蝶形弹簧，盘形弹簧
bending stress	弯曲应力
bevel gear	伞齿轮，锥齿轮
billet n.	方坯，坯锭
bimetallic [baimiˈtælik] a.	双金属的

bioengineering [ˌbaiəuˌendʒiˈniəriŋ] n.	生物工程(学)
biomechanics [ˌbaiəuməˈkæniks] n.	生物力学
bionics [baiˈɔniks] n.	仿生学
bioprinting n.	生物打印
blackout n.	断电;灯火熄灭
blade [bleid] n.	(机器上旋转的)叶片,桨叶
blank n.	板坯,坯料
blank holder	压边圈,压料板;坯缘压牢器
boiler [ˈbɔilə] n.	锅炉
bolt [bəult] n.	螺栓
bolt cutter	螺栓割刀;断线钳
bonded mica flake	屏蔽云母片
boon [buːn] n.	非常有用的东西;恩惠,益处
bore n. & v.	镗孔,内径 & 钻孔,镗孔
bow [bəu] n.	弓;船头
bracket [ˈbrækit] n.	支架,托架
brake fluid	制动液,刹车油
brine [brain] n.	盐水
brittleness n.	脆性,脆度
brush n.	电刷
brush-holder n.	电刷支架
bulk density	体积密度,松装密度,堆积密度
bulk supply	大批量供应
bus bar	汇流排,母线
bushing n.	轴衬,套管,衬套

C

cam [kæm] n.	凸轮
can opener	开罐器
cantilever [ˈkæntiliːvə] spring	悬臂弹簧
capacitor [kəˈpæsitə] n.	电容器
capital cost	基建费,投资费
capitalize v.	利用;积累资本;首字母大写
carbide [kɑːbaid] n.	碳化物
carbon footprint	碳足迹
carbonnanotube	碳纳米管

carbon soot	炭黑
carbon strip	碳棒
carburization [ˌkɑːbjuraiˈzeiʃən] n.	表面渗碳法
carrier n.	载波；载体
case hardening	表面硬化，表面淬火
cast ingot	铸锭
cast iron	铸铁；生铁
casting n. & v.	铸造；铸件 & 浇铸
casting to shape	铸型
catalytic [ˌkætəˈlitik] a.	促进性的；起催化作用的
cathode [ˈkæθəud] n.	阴极
cavity [ˈkæviti] n.	空腔；沟槽
cementite [siˈmentait] n.	渗碳体；碳化铁
centrifugal [ˌsentriˈfjuːgl] casting	离心式铸造
centrifuging n.	离心法
ceramics [səˈræmiks] n.	陶瓷
chain drive	链条传动
change gear	换挡
chaplet [ˈtʃæplət] n.	型芯撑
charcoal [ˈtʃɑːkəul] n.	木炭；活性炭
charger n.	充电器
chisel [ˈtʃizəl] n.	凿子
chlorine [kləːriːn] n.	氯
chromium [ˈkrəumiəm] n.	铬
circuitry [ˈsəːkitri] n.	电路系统，电路装置
city refuse [ˈrefjuːs]	城市垃圾
clamp v.	夹紧，固定住
clamping flange	钳式法兰
clinical trial	临床试验
closed die forging	闭式模锻
closed loop	闭合回路
clutch [klʌtʃ] n.	离合器
coarse structure	粗糙结构，粗松组织
coating n.	涂层，镀膜
coaxial [kəuˈæksiəl] cable	同轴电缆
cobalt [ˈkəubɔːlt] n.	钴

cobot [kəubət] n.	协作式机器人,人机合作机器人
coefficient of friction	摩擦系数
coil [kɔil] n.	线圈,绕组
coil spring	螺旋弹簧
coke [kəuk] n.	焦炭
collaborate [kəˈlæbəreit] v.	合作,协作
collaborative [kəˈlæbərətiv] robot	协作机器人
collinear shaft	共线轴
colloidal [kəˈlɔidəl] graphite	石墨乳
commission [kəˈmiʃn] v.	调试;委托
communication hub	通讯集线器
commutating pole	整流极
commutation [ˌkɔmjuˈteiʃn] n.	换向
commutator [ˈkɔmjuteitə] n.	换向器
compass [ˈkʌmpəs] n.	指南针
complement [ˈkɔmplimənt] v.	补充,补足
component [kəmˈpəunənt] n.	组件,元件
composite [ˈkɔmpəzit] n.	复合材料
compound n.	化合物,混合物
compressive load	压缩载荷
condensation [ˌkɔndenˈseiʃn] n.	冷凝,凝结
condenser [kənˈdensə] n.	冷凝器,电容器
conduct v.	导电,导热
conductor n.	导体
conduit [ˈkɔndjuit] n.	管道
configuration [kənˌfigəˈreiʃən] n.	配置,结构
connecting rod	连杆
consortium [kənˈsɔːtiəm] n.	联盟;联营企业
constant n.	常数;恒量
continuous cast ingot	连续铸锭
control engineering	控制工程
convene [kənˈviːn] v.	聚集,集合;召集
converge [kənˈvəːdʒ] v.	聚集,汇集
conveyance [kənˈveiəns] n.	运输,输送
coordinate [kəuˈɔːdineit] n.	坐标
cope [kəup] n.	上型箱,上砂箱

copper segment	换向器铜片
core *n.*	型芯
corkscrew [ˈkɔːkskruː] *n.*	开瓶器,螺丝锥
corpuscle [ˈkɔːpʌsl] *n.*	微粒;细胞
corrosion resistance	耐腐蚀性,抗腐蚀性
counteract [ˌkaʊntərˈækt] *v.*	抵消;中和;阻碍
counterpart *n.*	对应的事物,配对物
coupling [ˈkʌplɪŋ] *n.*	联轴器
crank [kræŋk] *n.*	曲柄
crankshaft *n.*	曲轴,机轴
creep *n.*	蠕变(缓慢变形)
critical point	临界点
critical temperature	临界温度
cross belt	交叉皮带
cross-linking *n.*	交联
cross-section *n.*	横截面
crucible [kruːsɪbl] *n.*	坩埚
cruise control	定速巡航,巡航控制
cure *v.*	使硬化,凝固
cushion *v.*	起缓冲作用;(用垫子)使柔和
cutting-edge *n.* & *a.*	切削刃,刃口 & 领先的,尖端的
cybernetics [ˌsaɪbəˈnetɪks] *n.*	控制论
cylinder [ˈsɪlɪndə] *n.*	圆柱体;(发动机的)气缸
cylindrical [səˈlɪndrɪkl] *a.*	圆柱形的

D

debut [deɪˈbjuː] *n.*	初次登台;开张
deep drawing	深压成型,深拉
default state	缺省状态
deflection [dɪˈflekʃn] *n.*	挠曲;偏差
deformation [ˌdiːfɔːˈmeɪʃən] *n.*	变形
depletion [dɪˈpliːʃən] *v.*	耗尽,干枯,枯竭
deployment [dɪˈplɔɪmənt] *n.*	部署,调度
dereliction [ˌderəˈlɪkʃn] *n.*	玩忽职守;抛弃,遗弃
derived unit	导出单位
deteriorate [dɪˈtɪəriəreɪt] *v.*	变坏,恶化

deterioration [diːˌtiəriəˈreiʃən] n.	恶化,变坏
detrimental [ˌdetriˈmentl] a.	不利的,有害的
diamagnetism n.	抗磁性;反磁性
diameter [daiˈæmitə] n.	直径
die n.	模具,冲模,钢模
die casting	压铸;拉模铸造
diesel [ˈdiːzl] n.	柴油
diode [ˈdaiəud] n.	(电子)二极管
direct current	直流电
direct-chill cast ingot	直冷式铸锭
direction of feed	进刀方向,进给方向
discern [diˈsəːn] v.	分辨,觉察,识别
discharge tube	放电管
discrete [diˈskriːt] component	离散元件,分立元件
discrete quantities	离散量
dissipate [ˈdisipeit] v.	消散,驱散
distill [diˈstil] v.	提取;蒸馏
distribution n.	分配,供应;分布
domotics [dəuˈmɔtiks] n.	居家机器人;家庭自动化
draft tube	尾水管
drag [dræg] n.	下型箱,下砂箱
draught [drɑːft] n.	气流
drill n. & v.	钻子 & 钻孔
drill rod	钻杆
drudgery [ˈdrʌdʒəri] n.	苦差事;单调沉闷的工作
dubbed [dʌbd] a.	被称为的;译制的
ductility [dʌkˈtiliti] n.	延展性;柔软性;塑性
dynamic [daiˈnæmik] a.	动态的,动力的

E

eddy-current loss	涡流损耗
elasticity [ˌiːlæˈstisəti] n.	弹性,弹力
elastomer [iˈlæstəmə] n.	弹性体,弹性材料
electric circuit	电路
electric power system	电力系统
electric ray	电鳐

electrical charge	电荷
electrical power	电力;电源;电功率
electric-arc furnace	电弧炉
electrochemical property	电化学性能
electrode [iˈlektrəud] n.	电极
electrolysis [iˌlekˈtrɔləsis] n.	电解
electrolyte [iˈlektrəulait] n.	电解液,电解质
electrolytic [iˌlektrəˈlitik] process	电解过程
electromagnetic [iˌlektrəumægˈnetik] a.	电磁的
electromagnetic effect	电磁效应
electromagnetic levitation (maglev) train	磁悬浮列车
electromagnetism [iˌlektrəuˈmægnətizəm] n.	电磁,电磁学
electromotive force	电动势
electron [iˈlektrɔn] n.	电子
electron accelerator	电子加速器
electronic circuit	电子线路
electronic engineering	电子工程
electronics [iˌlekˈtrɔniks] n.	电子学
electroplate v.	电镀
elemental carbon	元素碳
embed v.	嵌入,内嵌
encompass [inˈkʌmpəs] v.	围绕,包围
envelop [inˈveləp] v.	包围;遮盖
epicyclic [ˌepiˈsaiklik] gear	行星齿轮
epoxy [iˈpɔksi] n.	环氧树脂
equilibrium [ˌi:kwiˈlibriəm] n.	平衡,均衡
even number	偶数
evenly spaced	均匀分布
exciting current	励磁电流
exotherm [ˈeksəuˌθə:m] n.	温升,放热
exponential [ˌekspəˈnenʃl] n. & a.	指数 & 指数的
extruder [eksˈtru:də] n.	(塑料)挤出机
extrusion [iksˈtru:ʒn] n.	挤压加工
eye n.	吊环,孔

F

fastener ['fɑːsənə] n.	固定器,紧固零件
fatigue [fə'tiːg] strength	疲劳强度
feed v.	进刀,进料,进给
ferromagnetic [ˌferəumæg'netik] a.	铁磁的
ferrous metal	黑色金属
field coil	励磁线圈
filament n.	长纤丝;灯丝
film n.	薄膜
fine grain	细晶粒
finish n. & v.	光洁度 & 精加工
fire brick	耐火砖,防火砖
fixture n.	固定装置,设备
flange coupling	凸缘联轴节
flash n.	飞边
flashing n.	防水板,遮雨板
flask n.	砂箱,上箱
flat spring	平板弹簧,片弹簧
flexible body	柔性体,弹性体
flexible coupling	弹性联轴器
flux [flʌks] n.	磁通量
follower n.	从动件,从动轮
forging n.	锻造;锻件
formidable ['fɔːmidəbl] a.	强大的,令人敬畏的
forming n.	成型,成型加工
fracture ['fræktʃə] v.	折断,断裂
frame [freim] n.	框架;车架
frequency n.	频率
fuel tank	燃油箱
fulcrum ['fuːlkrəm] n.	支点
full annealing	完全退火
full-load current	满负载电流,全负荷电流
furnace ['fəːnis] n.	熔炉,火炉
furnace lining	炉衬
fusibility [ˌfjusi'biliti] n.	可熔性,熔度

G

gain n.	(电)增益
gating system	浇铸系统
gear [giə] n. & v.	齿轮 & 适合
gear ratio	齿轮比
gear train	齿轮传动链,轮系
gearbox n.	变速箱,齿轮箱
gel [dʒel] n.	凝胶
generate ['dʒenəreit] v.	发电;产生,生成
generation [ˌdʒenə'reiʃn] v.	发电
generator ['dʒenəreitə] n.	发电机
generic [dʒə'nerik] a.	类属的;通用的
geometric [ˌdʒi:ə'metrik] a.	几何学的
geometry [dʒi'ɔmitri] n.	几何学
germanium [dʒə:'meiniəm] n.	锗
give way to	为……所替代,让位于
gout [gaut] n.	痛风(病)
grain n.	晶粒,纹理
grain boundary	晶界
graphite ['græfait] n.	石墨
graphitized ['græfəˌtaizd] a.	石墨化的
grid [grid] n.	输电网;网格
grind [graind] v.	磨削,研磨
groove [gru:v] n.	槽
gum [gʌm] n.	黏胶,树胶

H

hammer head	锤头,榔头
hardening n.	硬化,淬火
hardness penetration	淬硬深度,淬火深度,硬化深度
harness ['hɑ:nis] v.	利用;治理
havoc ['hævək] n.	大破坏,灾害
head of water	水头
heart-lung n.	人工心肺机
heat treatment	热处理

heavy-duty *a*.	重型的,重负荷的;耐用的
helical gear	斜齿轮
helical spring	螺旋弹簧
helix ['hi:liks] *n*.	螺旋,螺旋状物
herald ['herəld] *v*.	预示……的来临;公开称赞
hertz [hə:ts] *n*.	赫兹(声波频率单位)
hexagonal [hek'sægənl] *a*.	六边的,六角形的
high voltage	高压
high-end *a*.	高端的,高档的
hinge [hindʒ] *n*.	铰链
horseshoe magnet	蹄(U)形磁铁
hot working	热加工,热作
household appliance	家用电器,家电
hub [hʌb] *n*.	中心,毂
hydraulic [hai'drɔlik] *a*.	液压的,水力的
hydroelectric station	水力发电站,水电站
hydrogen ['haidrədʒən] *n*.	氢
hydrologic [ˌhaidrə'lɔdʒikəl] *a*.	水文的
hype [haip] *n*.	大肆宣传
hypersonic [ˌhaipə'sɔnik] *a*.	特超音速的
hysteresis [ˌhistə'ri:sis] loss	磁滞损耗

I

impact strength	冲击强度
impart [im'pɑ:t] *v*.	给予,赋予
impetus ['impitəs] *n*.	推动力,冲力;促进
impulse turbine	冲击式水轮机
impurity *n*.	不纯,杂质
in series	串联
incandescent [ˌinkən'desnt] light	白炽灯
incentive [in'sentiv] *n*.	激励,刺激
inclined plane	斜面,倾斜面
indentation [ˌinden'teiʃən] *n*.	压痕,刻痕
inductor [in'dʌktə] *n*.	电感器,感应器
inertial [i'nə:ʃl] *a*.	惯性的
inertial guidance system	惯性导航系统,惯性制导系统

inexhaustible [ˌinigˈzɔːstəbl] a.	用之不竭的
infrastructure [ˈinfrəstrʌktʃə] n.	基础设施;下部构造
ingenious [inˈdʒiːniəs] a.	精巧的,巧妙的,新颖独特的
ingot [ˈiŋgət] n.	铸块,锭
ingot casting	铸锭,钢锭浇铸
instrumentation engineering	仪表工程
insulator [ˈinsəleitə] n.	绝缘体
integral [ˈintigrəl] a.	完整的,不可或缺的
integrated circuit	集成电路
intellectual property	知识产权
intermediate annealing	中间退火
internal cavity	内腔
internal combustion engine	内燃机
internal strain	内应变
internal stress	内应力
interplanetary spaceship	星际飞船
interpole [ˈintəpəul] n.	换向极
intervention [ˌintəˈvenʃn] n.	干涉,干预
intriguing [inˈtriːgiŋ] a.	非常有趣的,引人入胜的
inventory [ˈinvəntɔːri] n. & v.	清单;存货 & 开列清单
inversely [ˌinˈvəːsli] adv.	成反比地,相反地
inverter [inˈvəːtə] n.	逆变器
investment casting	熔模铸造,精密铸造
ion [ˈaiən] n.	离子
ionize [ˈaiənaiz] v.	(使)电离,(使)成离子
iron ore	铁矿
irrigate [ˈirigeit] v.	灌溉

J

jerk [dʒəːk] n.	急拉,急推
joint n.	关节,接合点
jolt [dʒəult] n. & v.	震动,颠簸,摇晃
journal bearing	滑动轴承;轴颈轴承

K

kilovolt [ˈkiləuvəult] n.	千伏

kinematic [ˌkinəˈmætik] a. 运动的；运动学的
kinetic [kiˈnetik] energy 动能

L

ladle [ˈleidl] v. 把……倒入，用长柄勺舀
laminated iron core 迭片式铁芯
lamination [ˌlæmiˈneiʃən] n. 迭片结构，层压
lead peroxide [pəˈrɔksaid] 过氧化铅
leaf spring 钢板弹簧（简称板簧）；弹簧片
leakage flux 渗漏通量
lean production 精益生产
levelized-cost-of-energy 能源平准化成本
lever [ˈliːvə] n. 杠杆
Leyden [ˈlaidn] jar 莱顿瓶
liability [ˌlaiəˈbiləti] n. 责任，义务
light microscope 光学显微镜
line of force 力线
line shaft 传动轴
linear [ˈliniə] a. 直线的，线性的
link n. 连杆
linkage [ˈliŋkidʒ] n. 联动装置，连杆机构；连接
lithium-ion [ˈliθiəm ˈaiən] n. 锂离子
living tissue [ˈtiʃuː] 生物组织，活组织
load n. 负载，负荷
low-flow turbine 低流量涡轮机
low-head hydropower 低水头水力发电
lubricant [ˈluːbrikənt] n. 润滑油，润滑剂
lubrication [ˌluːbriˈkeiʃn] n. 润滑，润滑作用
luminous [ˈluːminəs] a. 发光的，发亮的

M

machine element 机械零件，机械元件
magnesium [mæɡˈniːziəm] n. 镁
magnet [ˈmæɡnət] n. 磁铁，磁石
magnetic flux pinning 磁通钉扎
magnetic reluctance 磁阻

magnetism [ˈmægnətizəm] n.	磁性,磁力;吸引力
magnetomotive force	磁动势,磁通势
maintenance [ˈmeintənəns] n.	维护,保养;保持
malleability [ˌmæliəˈbiləti] n.	展延性;可锻性
malleable [ˈmæliəbəl] a.	有展延性的;可锻的
manganese [ˈmæŋgəniːz] n.	锰
manual transmission	手动变速箱
mass damper	质量阻尼器
mating surface	啮合面,配合面
mechanical advantage	机械效益
mechanism [ˈmekənizəm] n.	机制,机械装置
mechatronics [ˌmekəˈtrɔniks] n.	机械电子学,机电一体化
megawatt [ˈmegəwɔt] n.	兆瓦,百万瓦特
mercury [ˈməːkjəri] n.	水银,汞
mesh v.	啮合
metallurgical [ˌmetəˈləːdʒikl] a.	冶金的,冶金学的
metallurgist [məˈtælədʒist] n.	冶金家
metropolitan [ˌmetrəˈpɔlitən] a.	大城市的,大都会的
microscopic [ˌmaikrəˈskɔpik] a.	极小的,微小的
misalignment [ˌmisəˈlainmənt] n.	未对准;移位
miter gear	等径伞齿轮;斜方齿轮
mitigate [ˈmitigeit] v.	使缓和,使减轻
modulation [ˌmɔdjuˈleiʃən] n.	调制,调谐
molecular [məˈlekjələ] engineering	分子工程学
molecular magnet	分子磁铁
molecule [ˈmɔlikjuːl] n.	分子
molybdenum [məˈlibdənəm] n.	钼
momentous [məuˈmentəs] a.	重大的,重要的
motor [ˈməutə] n.	电机,电动机,马达
municipality [mjuːˌnisiˈpæləti] n.	自治市
mutual induction	互感

N

nanometer [ˈnænəumiːtə] n. (abbr. nm)	纳米(即十亿分之一米)
nanotechnology [ˌnænəutekˈnɔlədʒi] n.	纳米技术
navigational [ˌnæviˈgeiʃnl] a.	航行的,航运的,导航的

needle bearing	滚针轴承
negate [ni'geit] v.	否定，取消
net charge	净电荷
neutron ['nju:trɔn] n.	中子
nickel ['nikl] n.	镍
nodular ['nɔdʒələ] iron	球墨铸铁
no-load operation	空载运行
nonferrous metal	有色金属
normalizing ['nɔ:məlaiziŋ] n.	正火
notoriety [ˌnəutə'raiəti] n.	恶名，坏名声
nozzle ['nɔzl] n.	喷嘴
nuclear fusion ['fju:ʒn]	核子融合
nuclei ['nju:kliai] n.	原子核（nucleus 的复数形式）
nucleus ['nju:kliəs] n.	原子核
numbing ['nʌmiŋ] a.	令人麻木的，使人失去知觉的
nut [nʌt] n.	螺母，螺帽

O

omnipresent [ˌɔmni'preznt] a.	无所不在的，遍及各处的
open belt	开口皮带
open die forging	开式模锻
open riser	明冒口
open source	开放源码，开源
operation research	运筹学
optical fiber	光纤，光导纤维
optimum ['ɔptiməm] a.	最佳的，最适宜的
ore [ɔ:] n.	矿石，矿砂
outlet n.	电源插座
oval-shaped arc	椭圆形弧线
overhaul ['əuvəhɔ:l] n. & v.	全面改革；彻底检修

P

pandemic [pæn'demik] n.	流行病，瘟疫
panoramic [ˌpænə'ræmik] a.	全景的
paperclip n.	回形针，曲别针
paradoxical [ˌpærə'dɔksikl] a.	似自相矛盾的

parting line	分型线,分模线
pearlite ['pəːlait] n.	珠光体
pedal ['pedl] v. & n.	踩踏板,骑车 & 踏板
penstock n.	压力管道
performance n.	性能
periphery [pə'rifəri] n.	边缘,周围
permanent-mold casting	永久型铸造,恒模铸造
pesky ['peski] a.	讨厌的,麻烦的
phosphorus ['fɔsfərəs] n.	磷
photoelectric cell	光电池,光电管
photoelectricity n.	光电
piezoelectricity [paiiːzəuilek'trisiti] n.	压电
pilot-plant stage	中试阶段
pinion ['pinjən] n.	小齿轮
pin-jointed a.	销接的
piston ['pistən] n.	活塞
piston pin	活塞销
pivot ['pivət] n. & v.	支点,枢轴 & 在枢轴上旋转
pivotal ['pivətl] a.	关键的,核心的
plain bearing	滑动轴承;平面轴承
plain carbon steel	普通碳钢,碳素钢
planetary electron	轨道电子
planetary gear	行星齿轮机构
plaster ['plɑːstə] n.	石膏;灰泥
plastic deformation	塑性变形
plasticity [plæ'stisəti] n.	塑性,可塑性
plate n.	金属板,厚钢板,板材
pliable ['plaiəbl] a.	易弯曲的,可塑的,柔韧的
plumber ['plʌmə] n.	管道工
plutonium [pluː'təuniəm] n.	钚(放射性化学元素)
pneumatic [njuː'mætik] a.	气动的,风动的
pole face	极面
pole shoe	极靴,极端
polish v.	抛光
polyester [ˌpɔli'estə] n.	聚酯纤维
polymer ['pɔlimə] n.	聚合物,聚合体

polymerize [ˈpɔliməraiz] v.	使聚合
portfolio [pɔːtˈfəuliəu] n.	系列计划；文件夹
pouring basin	外浇口，浇口杯
power engineering	电力工程；动力工程
power loss	功率损耗
power surge	电力高峰
power system	电力系统
powerhouse n.	动力源
powertrain n.	动力系统，传动系统
preservative treatment	防腐处理
pressure or reaction turbine	压力或反应式水轮机
presuppose [ˌpriːsəˈpəuz] v.	以……为先决条件；假定，预料
primary coil	初级线圈
prime mover	原动力
prismatic [prizˈmætik] joint	移动关节
proliferate [prəˈlifəreit] v.	激增，增殖
prominence [ˈprɔminəns] n.	显著，突出，卓越
proportionately [prəˈpɔːʃənətli] adv.	成比例地
proprietary [prəˈpraiətri] a.	专利的，专营的；所有权的
propulsion [prəˈpʌlʃn] n.	推动力，推进
pros and cons	正反两方面；有利有弊；赞成与反对
protocol [ˈprəutəkɔl] n.	（数据传递的）协议，规约
proton [ˈprəutɔn] n.	质子
prototype [ˈprəutətaip] n.	原型
province n.	（学问、活动或责任的）范围
pulley [ˈpuli] n.	滑轮，皮带轮
pulsar [ˈpʌlsɑː] n.	脉冲星
pumped-storage hydropower	抽水蓄能水力发电
punch [ˈpʌntʃ] n.	冲头，冲压机
pundit [ˈpʌndit] n.	权威，专家，行家
pylon [ˈpailən] n.	电缆塔，高压线铁塔

Q

quantum [ˈkwɔntəm] n.	量子
quantum levitation	量子悬浮
quantum locking	量子锁定

quarter-turn belt	直角回转皮带,直角挂轮皮带
quasar [ˈkweizɑː] n.	类星体;类星射电源
quenching [ˈkwentʃiŋ] n.	淬火;骤冷
quintessential [ˌkwintiˈsenʃl] a.	典型的,本质的,精髓的

R

raceway n.	滚道
rack n.	齿条
rack-and-pinion gear	齿轮齿条式齿轮
radial load	径向载荷
radio telescope	射电望远镜
radioactive [ˌreidiəuˈæktiv] a.	放射性的
radius [ˈreidiəs] n.	半径
rail n.	钢轨,轨条
ram n.	冲头
raw material	原材料
reactance [riˈæktəns] n.	电抗
rebound clip	弹簧箍圈,弹簧连接夹
reciprocal [riˈsiprəkl] n.	倒数
reciprocating [riˈsiprəˌkeitiŋ] a.	往复的
recrystallization [riːkristəlaiˈzeiʃən] n.	再结晶,重结晶
recrystallize [riːˈkristəlaiz] v.	再结晶,重结晶
rectangular [rekˈtæŋgjələ] a.	矩形的,长方形的
rectifier [ˈrektifaiə] n.	整流器
reduction [riˈdʌkʃn] n.	减少
refinement n.	改进,改善
refractory [riˈfræktəri] a. & n.	难熔的 & 耐火物质
regenerative braking	再生制动
repulsion [riˈpʌlʃn] n.	排斥,反感
reservoir [ˈrezəvwɑː] n.	水库,蓄水池
resin [ˈrezin] n.	树脂
resistor [riˈzistə] n.	电阻器
retina [ˈretinə] n.	视网膜
revamp [ˌriːˈvæmp] v.	改变,修改;翻新
revenue [ˈrevənjuː] n.	财政收入,税收收入,收益
revolute joint	转动关节

rheocasting [ˈriːəukæstiŋ] n.	流变铸造
rigid body	刚性体,刚体
rigid coupling	刚性联轴器;固定耦合
rivet heading	铆钉打头
rod steel	圆钢,钢杆
rolled steel	轧钢
roller n.	轧辊;滚轴
roller bearing	滚柱轴承
rolling n.	轧制
rolling bearing	滚动轴承
root mean square	均方根
rotary [ˈrəutəri] a.	旋转的,绕轴转动的
rotor [ˈrəutə] n.	转子,转动部件
round off	完成,结束
runner n.	流道,分流道
rupture [rʌptʃə] v.	破裂,断裂

S

sand casting	砂型铸造
sanitation [ˌsæniˈteiʃn] n.	环境卫生;卫生设备
satirical [səˈtirikl] a.	讽刺的,讥讽的
saturate [ˈsætʃəreit] v.	浸透;使饱和
scale v.	生成氧化皮
scarcity [ˈskeəsəti] n.	缺乏,不足
scenario [səˈnɑːriəu] n.	设想,方案
schematic [skiːˈmætik] a.	略图的,图解的
scratch [skrætʃ] v.	划损,刮坏
screw [skruː] n.	螺丝,螺钉,螺杆
screw joint	螺纹接头,螺旋接头
screwdriver [ˈskruːdraivə] n.	螺丝刀
seal n.	密封垫
secondary coil	次级线圈
self induction	自感,自感现象
self-excited a.	自励磁的,自激的
semicentrifugal casting	半离心式铸造
semiconductor n.	半导体

semi-continuous cast ingot	半连续性铸锭
sensor ['sensə] n.	传感器,敏感元件
separately excited	他励磁的,分激的
separator n.	挡板,隔板;轴承座
serpentine ['səːpəntain] belt	蛇纹岩带
set v.	凝固,凝结
shaft [ʃɑːft] n.	主轴
shatter ['ʃætə] v.	破碎,粉碎
shearing stress	剪应力
sheet n.	薄钢板,薄板
shock protection	防震保护
shrinkage ['ʃriŋkidʒ] n.	收缩
silicon ['silikən] n.	硅
silicone ['silikəun] n.	硅树脂,硅酮
simulation [ˌsimjuˈleiʃn] n.	模拟,仿真
single-phase system	单相系统
skyscraper n.	摩天大楼
slab n.	板坯
slider-crank n.	曲柄滑块
sliding bearing	滑动轴承
sliding joint	滑动关节
slot [slɔt] n.	槽
slurry ['sləːri] n.	泥浆
socket ['sɔkit] n.	插座
sodium ['səudiəm] n.	钠
solder [səuldə] n. & v.	焊料 & 焊接
soldering iron	烙铁,焊铁
solicitation [səˌlisiˈteiʃn] n.	(意见的)征求
solidification [səˌlidifiˈkeiʃn] n.	凝固,固化
solution n.	溶液
spare-part n.	零部件
spectral ['spektrəl] imaging	光谱成像
spectrum ['spektrəm] n.	范围;色谱,光谱
spherical ['sferikl] joint	球面关节
spheroidized [ˌsfiərɔiˈdaizd] annealing	球化退火
spider n.	支架,三脚架

Glossary

spiral ['spairəl] a.	螺旋形的
spline [splain] n.	花键
spontaneous [spɔn'teiniəs] a.	自发的,自然的
spring pressure	弹力
sprocket ['sprɔkit] n.	链轮齿,扣链齿轮
sprue [spruː] n.	注入口,浇口
spur gear	正齿轮,直齿轮
standing loss	标准损耗
static cast ingot	静态铸锭
stator ['steitə] n.	定子
steam jet	蒸汽喷射流
steam turbine	蒸汽涡轮
steering mechanism	转向机制,转向装置
stellar ['stelə] a.	出色的,精彩的;恒星的
step-down transformer	降压变压器
step-up transformer	升压变压器
stereolithography ['steriəuˌli'θɔgrəfi] n.	立体光刻,立体平版印刷
stiffness n.	刚度,挺度
strain [strein] n.	应变
strain hardening	应变硬化,加工硬化
stress-relief annealing	消除应力退火,去应力退火
structural steel	结构钢
submarine ['sʌbməriːn] a.	水下的,海底的
substation ['sʌbsteiʃn] n.	变电站
sulfuric [sʌl'fjuərik] acid	硫酸
sulphur ['sʌlfə] n.	硫黄
sun block	防晒霜
sun screen	隔离霜,防晒液
supply chain	供应链
surface finish	表面抛光
susceptibility [səˌseptə'biləti] n.	敏感性,灵敏度
susceptible [sə'septəbl] a.	易受影响的
suspension [sə'spenʃən] n.	汽车悬架,悬挂
symmetry ['simətri] n.	对称;相仿
synthesize ['sinθəsaiz] v.	合成,综合
synthetic [sin'θetik] rubber	合成橡胶

syringe [siˈrindʒ] n.	喷射器；注射器；吸管

T

telecommunications engineering	通信工程，电信工程
tempering [ˈtempəriŋ] n.	回火
tensile [ˈtensail] load	拉伸载荷
terminal [ˈtəːminl] n.	极，端子，终端
thermal aging	热老化
thermocouple n.	热电偶
thermodynamics [ˌθəːməudaiˈnæmiks] n.	热力学
thermoelectricity [ˈθəːməuiˌlekˈtrisəti] n.	热电
thermoplastic material	热塑性材料
thermosetting [ˈθəːməusetiŋ] material	热固性材料
three-phase system	三相系统
thrust load	轴向载荷，推力载荷
tolerance n.	公差
tong [ˈtɔŋ] n.	钳子
tooth n.	轮齿
torque [tɔːk] v.	扭矩，转矩
torsion load	扭力载荷
toughness n.	韧性，韧度
trace n.	痕量
traction [ˈtrækʃn] n.	牵引，拖拉
transceiver [trænˈsiːvə] n.	无线电收发两用机
transducer [trænzˈdjuːsə] n.	换能器，转换器
transformer [trænsˈfɔːmə] n.	变压器
transistor [trænˈsistə] n.	晶体管
transit [ˈtrænzit] system	公交系统；转接系统
translation n.	平移
transmission [trænsˈmiʃn] n.	传送，传递；变速器
transmission system	传动系统；传输系统
transmit [trænzˈmit] v.	传输，传送，传递
transmitter [trænzˈmitə] n.	发送器；发射机
transparency [trænsˈpærənsi] n.	透明，透明度
treated fabric	处理过的纤维
tremor [ˈtremə] n.	轻微地震，小震

trial and error	试错法
true centrifugal casting	真正的离心式铸造
tube drawing	管材拉拔,拔管
tuned circuit	调谐电路
tungsten ['tʌŋstən] n.	钨
turbine ['tə:bain] n.	涡轮机,汽轮机
turbine runner	涡轮机转轮
turnkey solution	整体解决方案,交钥匙解决方案
twitch [twitʃ] v.	抽动,颤动

U

ubiquitous [ju:'bikwitəs] a.	普遍存在的,无所不在的
ultimate strength	极限强度
unfathomable [ʌn'fæðəməbl] a.	难以理解的,高深莫测的
uranium [ju'reiniəm] n.	铀(放射性化学元素)
urethane ['jurəθein] n.	尿烷
utility-scale grid	公用事业规模输电网

V

vacuum ['vækju:m] n.	真空
value engineering	价值工程,工程经济学
valve [vælv] n.	阀门
vanadium [və'neidiəm] n.	钒
vane [vein] n.	叶片,轮叶
vaporization [ˌveipərai'zeiʃn] n.	蒸发;汽化
variable ['veəriəbl] n. & a.	变量 & 可变的
velocity [və'lɔsəti] n.	速度,速率
vendor ['vendə] n.	供应商
ventcap n.	透气盖
ventilate ['ventileit] v.	使通风
viability [ˌvaiə'biləti] n.	可行性
vibration [vai'breiʃn] n.	振动
vibration damping	减振
vice [vais] n.	老虎钳
viz.	即,也就是
Voltaic [və'teiik] cell	伏打电池

vortex ['vɔːteks] *n. pl.* vortices	低涡,涡旋

W

wafer ['weifə] *n.*	晶圆
warehouse ['weəhaus] *n.*	仓库,货仓
warp ['wɔːp] *v.*	弯曲,变形
weapons of mass destruction	大规模杀伤性武器
wear resistance	耐磨性,耐磨度
wedge [wedʒ] *n.*	楔子,楔形物
whirling ['wəːliŋ] *a.*	旋转的
winding ['waindiŋ] *n.*	绕组
winding mechanism	绕线机制,发条
windmill *n.*	风车;风车磨坊
wire band	线圈
wire drawing	拉线,拉丝
wire spring	钢丝弹簧
wobble ['wɔbl] *n. & v.*	摇晃
workflow *n.*	工作流程
workpiece *n.*	工件
worm *n.*	蜗杆
worm gear	蜗轮
wrought [rɔːt] iron	锻铁;熟铁

Y

yield point	屈服点
yoke [jəuk] *n.*	轭铁

Z

zinc [ziŋk] *n.*	锌

参 考 文 献

[1] Chen, W. K. The Electrical Engineering Handbook [M]. California: Elsevier Academic Press, 2004.

[2] Dixit, U. S, Hazarika, M. & Davim, J. P. A Brief History of Mechanical Engineering [M]. New York: Springer Publishing Company, 2017.

[3] Hatim, B. & Mason, I. Discourse and the Translator [M]. London: Longman, 1990.

[4] Johnson, D. Fundamentals of Electrical Engineering [M]. Houston: Rice University, 2008.

[5] Kumar, V. Analysis of Simple Planar Linkages [M]. Pennsylvania: University of Pennsylvania, 2006.

[6] Macaulay, D. The Way Things Work [M]. Boston: Houghton Mifflin, 1998.

[7] Mulukutla, S. The Oxford Series in Electrical and Computer Engineering [M]. London: Oxford University Press, 2000.

[8] Seleskovitch, D. & Ledere, M. Pedagogie Raisonnee de L' Interpretation [M]. Paris: Didier Erudition, 1989.

[9] Singh, R. Introduction to Basic Manufacturing Processes and Workshop Technology [M]. New Delhi: New Age International Ltd., Publishers, 2006.

[10] Trimble, L. English for Science and Technology [M]. Cambridge: Cambridge University Press, 1985.

[11] U. S. Bureau of Naval Personnel. Basic Machines and How They Work [M]. New York: Dover Publications, 1971.

[12] Widdowson, H. G. Teaching Language as Communication [M]. Oxford: Oxford University Press, 1978.

[13] Wickert, J. & Lewis, K. An Introduction to Mechanical Engineering (3rd edition) [M]. Connecticut: Cengage Learning, 2013.

[14] 卜玉坤. 机械英语[M]. 北京:外语教学与研究出版社,2001.

[15] 卜玉坤. 电气与电子英语[M]. 北京:外语教学与研究出版社,2002.

[16] 常晨光,陈瑜敏. 功能语境研究[M]. 北京:外语教学与研究出版社,2011.

[17] 高巍,范波. 科技英语翻译教学再思考:理论、途径和方法[J]. 外语电化教学,2020(5):65-71.

[18] 高巍,高颖超,陈燕如. 语篇功能视阈下科技英语段落与翻译研究[M]. 大连大学学报,2016(5):66-71.

[19] 梁志华. 科技英语的特点及其翻译策略[J]. 重庆交通大学学报(社科版),2009,9(3):125-128.

[20] 刘源甫. 科技翻译词义的具体化与抽象化引申[J]. 中国科技翻译,2004,17(2):16-19.

[21] 罗天. 新时代英汉翻译教程[M]. 北京:人民交通出版社股份有限公司,2019.

[22] 王大维,王中立,王军. 新时代科技英语翻译路径研究 [M]. 南京:东南大学出版社,2019.

[23] 张美芳. 功能途径论翻译:以英汉翻译为例[M]. 北京:外文出版社,2015.

[24] 张敏,杨秀芬. 科技英语阅读教程[M]. 北京:外语教学与研究出版社,2007.

[25] 张彦. 科学术语翻译概论[M]. 杭州:杭州大学出版社,2008.

[26] 张周易,毛明勇. 研究生实用英语教程[M]. 北京:人民交通出版社股份有限公司,2017.

[27] 邹晶明,欧阳美和. 机械英语[M]. 北京:高等教育出版社,2010.

[28] 左尚鸿. 钒钛科技英语翻译标准与翻译路径研究[J]. 外语教育研究,2017,5(4):60-64.

参 考 答 案

Unit 1

Section A

Ⅰ. 1. T 2. F 3. F 4. T 5. F

Ⅱ. 1. scientific discovery, product applications
   2. perceived social needs, commercial applications
   3. the quality of healthcare, the safety of food products, the operation of financial systems
   4. chemical engineering, civil engineering, electrical engineering, mechanical engineering
   5. Electrical engineering

Ⅲ. 1. Engineering is a driver of social and economic growth and an integral part of the business cycle.
   2. Civil engineers design, plan, and supervise the construction of buildings, highways, and transit systems.

Ⅳ.

English	Chinese
molecular engineering	分子工程学
transit system	公交系统
atom	原子
nanotechnology	纳米技术
trial and error	试错法
mechatronics	机械电子学
simulation	模拟,仿真

Ⅴ. 1. ingenious 2. integral 3. link 4. performance 5. prominence

Section B

Ⅰ. 1. C 2. C 3. A 4. B 5. B 6. D

Ⅱ. 1. generation, transmission, distribution
   2. design, maintenance
   3. Cruise
   4. radio engineering

5. communication channel

6. amplitude modulation, frequency modulation

Ⅲ. 1. Electrical engineering is an engineering discipline concerned with the study, design, and application of equipment, devices, and systems which use electricity, electronics, and electromagnetism.

2. The first example is the flight and propulsion systems of commercial airliners; the second example is the cruise control present in many modern automobiles.

3. Integrated circuits packed a large number—often millions—of tiny electrical components, mainly transistors, into a small chip around the size of a coin. This allowed for the powerful computers and other electronic devices we see today.

4. Instrumentation engineering deals with the design of devices to measure physical quantities such as pressure, flow, and temperature.

Ⅳ.

English	Chinese
amplitude modulation	振幅调制
cruise control	定速巡航,巡航控制
capacitor	电容器
component	组件,元件
integrated circuit	集成电路
power surge	电力高峰
household appliance	家用电器,家电
maintenance	维护,保养
variable	变量

Ⅴ. 1. e 2. d 3. a 4. g 5. b 6. f 7. c

Ⅵ. 电力工程涉及发电、输电和配电以及一系列相关设备的设计。这些设备包括变压器、发电机、电动机、高压工程和电力电子设备。在世界上许多地区,政府维护一个被称为电网的电力网络,该网络将各种发电机与其能源用户连接在一起。用户从电网购买电能,避免了自行发电的昂贵操作。电力工程师可以从事电网以及与其相连的电力系统的设计和维护工作。这类系统被称为并网电力系统,可以向电网提供额外电力,或从电网获取电力,亦或两者兼而有之。电力工程师也可能在不与电网相连的系统上工作,我们将其称为离网电力系统,在某些情况下,离网电力系统比并网系统更受青睐。未来的发展包括卫星控制电力系统,它能实时进行反馈以防止电力高峰和停电断电的发生。

Section C

Ⅰ. 1. B 2. C 3. B 4. D 5. A

II. 1. machines, the production of power
2. economically, minimized
3. the control system
4. an ample supply of power
5. Jet aircraft, nuclear reactors

III. 1. The fields of mechanical engineering are: 1) the development of machines for the production of goods; 2) the development of machines for the production of power; 3) the development of military weapons; 4) bioengineering; and 5) environmental control.
2. A completely automated machine shop is for batch production, operating on a three-shift basis but attended by a staff for only one shift per day.
3. Bioengineering is a relatively new and distinct field of mechanical engineering that includes the provision of machines to replace or augment the functions of the human body and of equipment for use in medical treatment.
4. The four functions of mechanical engineering are as follows: 1) the understanding of and dealing with the bases of mechanical science; 2) the sequence of research, design, and development; 3) the production of products and power, which embraces planning, operation, and maintenance; and 4) the coordinating function of the mechanical engineering, including management, consulting, and, in some cases, marketing.
5. The threats include the exponential increase in population, the environmental pollution and the exhaustion of natural resources which will seriously affect people's living standards.

IV.

English	Chinese
bioengineering	生物工程
exponential	指数
lubrication	润滑
spare-part	零部件
viability	可行性
batch production	批量生产
capital cost	基建费,投资费
thermodynamics	热力学

V. 1. impetus 2. accuracy 3. province 4. collaborate 5. maximum

Ⅵ. 1. 机械工程已经由一门主要基于试错法的技工应用的技艺发展成为职业工程师在研究、设计和生产领域使用的科学方法。

2. 大型电厂和整个核电站的控制系统已变成高度复杂的电子、流体、电、水力和机械零件的网络,这一切都涉及机械工程师的所有学术领域。

3. 然而,创造力是无法合理化的。正如在其他领域一样,在机械工程中,能够采取重要的、出人意料的,并能开创出新方法的能力,仍然具有个人的、即兴的特点。

Unit 2

Section A

Ⅰ. 1. T 2. F 3. F 4. T 5. T

Ⅱ. 1. Thunderer of the Nile
   2. electrical
   3. producing a simple form of cell
   4. electricity, magnetism
   5. electromagnetic effects

Ⅲ. 1. A Leyden jar consisted merely of a bottle containing water (or alcohol or mercury) with a long nail dipping into the liquid.

2. Because it meant that for the first time it was possible to maintain an electric current for a long period.

3. Oersted discovered that there is a relationship between electricity and magnetism. He found out that when current passed through the wire, the compass needle was moved.

Ⅳ.

English	Chinese
condenser	冷凝器
magnetism	磁性
electrical charge	电荷
electromagnetic effect	电磁效应
coil	线圈,绕组

Ⅴ. Lightning is one of the electric phenomena commonly seen in nature. People in the past knew little about lightning except for its horrible damage on earth. Franklin's famous experiment "kite in a thunderstorm" opened a new page of recognizing lightning. With deeper and deeper electricity study, people now know that lightning has something to do with the sudden great discharge in cumulonimbus. Now man is able to protect himself well from being struck by lightning.

Section B

Ⅰ. 1. B 2. D 3. A 4. D 5. C 6. D 7. A 8. D

Ⅱ. 1. electric charge

    2. repel, attract

    3. protons, neutrons

    4. planetary electrons, free electrons

    5. insulators

    6. negatively charged, positively charged

Ⅲ. 1. Some materials permit the movement of free electrons more easily than others. These materials are called conductors. Good examples of conductors are copper, silver, and aluminum.

    2. A battery is a group of cells connected together to produce a greater amount of electricity.

    3. The first problem is a chemical reaction that deposits hydrogen bubbles on the anode, preventing electrons from passing into the electrode. The second problem is that the cathode will gradually dissolve as a result of the continuous chemical reaction with the electrolyte.

    4. In storage batteries like those used in automobiles, the chemical reactions can be reversed by passing a reversed electric current through the electrolyte. This process is known as recharging, and it lengthens the life of the battery considerably.

Ⅳ. Generally speaking, an atom is composed of a nucleus and some electrons which orbit around the nucleus. And the nucleus itself consists of two particles, namely protons and neutrons. The number of protons equals that of electrons, which makes it possible to give each chemical element an atomic number. The nucleus is in the center of the atom and contains almost the entire weight or mass of it; while the electrons only have a very small mass. Electrons can be drawn away from their nucleus and become free electrons whose flow forms the electric current.

Ⅴ.

English	Chinese
aluminum	铝
conductor	导体
electrolyte	电解液,电解质
molecule	分子
electrochemical property	电化学性能
planetary electron	轨道电子

续上表

English	Chinese
nucleus	原子核
separator	隔板
atomic number	原子序数

Ⅵ. 1. e 2. a 3. d 4. b 5. f 6. c

Ⅶ. 1. 电子的移动需要能量,能量可以定义为做功的能力。电的流动需要大量的能量,这一点可以从异种电荷材料间迸发火星这一现象中看出。

2. 伏打电池包括两条置于盐溶液中的金属。其中一种的化学活性强于另一种,并且释放出电子,电子被活性较弱的金属所吸引。两条金属通过一根导线于外部连接而产生一条回路——电流的通路。

3. 汽车中使用的蓄电池,可让逆向电流通过电解液使化学反应逆行进行。该过程被称为充电,它可以很好地延长电池的使用寿命。

Section C

Ⅰ. 1. F 2. T 3. F 4. T 5. F 6. F

Ⅱ. 1. electric current, electromagnetic effects

2. the structure of the atom

3. Electronic refinements

4. rectifier

5. Thermoelectric generators

6. piezoelectricity, thermoelectricity, photoelectricity

Ⅲ. Briefly, there are three types of electricity: 1) direct current and alternating current electricity; 2) high-volt and low-volt electricity; 3) practicable electricity, such as firepower, waterpower, nuclear and impracticable electricity, such as piezoelectricity, thermoelectricity, photoelectricity.

Ⅳ.

English	Chinese
photoelectricity	光电
piezoelectricity	压电
semiconductor	半导体
turbine	涡轮机,汽轮机
alternating current	交流电
atomic fission	原子裂变
rectifier	整流器

续上表

English	Chinese
electromotive force	电动势
kinetic energy	动能

Ⅴ. 1. pioneering 2. remarkable 3. refer 4. transmit 5. exploit

Ⅵ. The first demonstration electric vehicle was made in the 1830s, and commercial electric vehicles were available by the end of the 19th century. However, electric vehicles didn't enjoy the enormous success of internal combustion engine vehicles, whose normal advantages include longer ranges and easy refueling. Today's concerns about the environment, particularly noise and exhaust emissions, coupled with new developments in batteries and fuel cells, gradually swing the balance back in favour of electric vehicles.

Unit 3

Section A

Ⅰ. 1. F 2. F 3. F 4. F 5. T

Ⅱ. 1. alternating current

 2. constant speed work

 3. flow of electricity, constant value

 4. the waveform of alternating current

 5. standardize frequencies

Ⅲ. 1. D. c. refers to direct current, and a.c. alternating current. In direct current, the flow of electrons moves steadily in one direction; and in alternating current, the flow of electrons is reversed rapidly over and over again.

 2. Because alternating current can be transformed to higher or lower voltages, thereby facilitating the transmission of high power over considerable lengths of line and reducing costs.

 3. Frequency is defined as the number of periods occurring in the unit of time. It is the reciprocal of time and can be expressed in hertz.

 4. 50 hertz.

Ⅳ.

English	Chinese
alternating current	交流电
direct current	直流电

续上表

English	Chinese
a function of time	时间函数
reciprocal	倒数
frequency	频率
constant	常数

Section B

Ⅰ. 1. D 2. B 3. A 4. B 5. C 6. C

Ⅱ. 1. three-phase system

   2. coal, oil, natural gas

   3. turbines

   4. large electricity generators

   5. impulse turbines, pressure or reaction turbines

   6. suspension-type insulators

Ⅲ. 1. Because the power industry manufactures its product at the very instant that it is required by the customer.

   2. The sources are steam obtained by means of a conventional fuel (coal, oil, or natural gas), the combustion of city refuse or the employment of nuclear fuel, water, and diesel power from oil.

   3. Because in condensing turbines, the condenser is able to increase the expansion ratio of the steam, the efficiency, and work output of the turbine.

   4. In conventional plants steam is generated by steam generators; while in nuclear plants it is generated with the aid of a reactor in which the controlled fission of uranium or plutonium provides the necessary heat for the vaporization of water.

   5. The diesel engine has the advantages of low fuel cost, a brief warming-up period, low standing losses, and little cooling water.

   6. In large metropolitan areas, both overhead and underground distribution systems are used. In smaller towns and in the less congested districts of large cities, overhead distribution system is usually adopted. This is because the government should take the cost into consideration.

Ⅳ. A turbine consists of a shaft and a rotor fixed in bearings and enclosed in a cylindrical casing, and the shaft of the turbine is directly connected to a large generator. When jets of steam from nozzles strike the blades attached to the shaft, the turbine rotates smoothly, synchronously causing the other shaft of the generator rotate in a magnetic field. When a conductor is moved through a magnetic field in such a way as to cut the magnetic lines, electricity is generated.

V.

English	Chinese
cylinder	气缸
substation	变电站
city refuse	城市垃圾
cross-section	横截面
automatic breaker	自动断路器
bulk supply	大批量供应
pilot-plant stage	中试阶段
hydroelectric station	水力发电站
three-phase system	三相系统
diesel engine	柴油发动机

Ⅵ. 1. b 2. f 3. d 4. g 5. e 6. c 7. a

Ⅶ. 1. 显而易见,对电力系统而言服务的连续性至关重要。没有哪种服务能够完全避免可能出现的失误,而系统的成本显然依赖于其稳定性。

2. 然而,网络可靠性的增加是通过应用一定数量的生成单元和在发电站汇流排各分区之间以及在国内、国际电网传输线路中使用自动断路器得以实现的。

3. 涡轮中的蒸汽具有能动性。蒸汽进入涡轮时压力较高、体积较小,而离开时却压力较低、体积较大。

4. 高压传输通常使用配以悬挂式绝缘设备的高架结构。称为电缆的铁塔用于承载绝缘设备,各导体悬挂在一组或一串绝缘体的底部。

Section C

Ⅰ. 1. T 2. T 3. T 4. F 5. T 6. F

Ⅱ. 1. Hydroelectric power

2. Pumped-storage hydropower

3. simple, robust, economical

4. cost reductions, water-use optimization, facility upgrades

5. Water-Use Optimization Toolset

Ⅲ. 1. The Water Power Program's work in hydropower technology development covers the areas of: 1) low-head hydropower; 2) materials and manufacturing; 3) hydropower systems; and 4) hydropower technology accomplishments.

2. The Water Power Program funds R&D to identify and test new materials and manufacturing techniques to improve the performance and lower the costs of

hydropower.

3. WUOT includes tools for hydrologic forecasting, seasonal hydro-systems analysis, day-ahead scheduling, real-time operations, and environmental performance operations.

Ⅳ.

English	Chinese
commission	调试
megawatt	兆瓦
powertrain	动力系统
turbine runner	涡轮机转轮
coating	涂层
pumped-storage hydropower	抽水蓄能水力发电

Ⅴ. 1. conveyance   2. ancillary   3. catalytic   4. deterioration   5. revamped

Ⅵ. Hydropower stations use hydraulic turbines that need to be supplied with water falling or flowing from a height above the turbine to develop shaft work. The pressure of water flowing to the turbine forces it through nozzles or blades at high speed. The curved blades make the steam jets change their direction and exert a force on the moving blades of the turbine. The blades are mounted on a wheel and a shaft. The rotation of the shaft drives the direct-connected electric generator. The spent water flows from the turbine to the river at a lower elevation.

Unit 4

Section A

Ⅰ. 1. T   2. F   3. F   4. T   5. T   6. F   7. F   8. T

Ⅱ. 1. more lines of force

2. Hans Oersted

3. commutator

4. stator, rotor

5. rotary, reciprocating

6. Transformers

Ⅲ. 1. For electricity, like electric charges repel and unlike charges attract; while for magnets, like poles repel and unlike poles attract.

2. Because if the poles of the armature reached a position directly opposite to the unlike poles of the horseshoe magnet, the armature would become locked and no further

261

motion would be possible.

3. As for a step-up transformer, the secondary coil has more turns than the primary coil; and as for a step-down transformer, the primary coil has more turns than the secondary coil. Step-up transformers are used at a power plant to increase the voltage for transmission; while step-down transformers are used to change the power to voltages suitable for industrial or domestic use.

Ⅳ.

English	Chinese
primary coil	初级线圈
pivot	支点,枢轴
closed loop	闭合回路
bar magnet	条形磁铁
armature	电枢
step-up transformer	升压变压器

Ⅴ. About 1800 A. D., it was noticed that a compass was affected when it was brought near a wire carrying current. Hans Christian Oersted, a Danish engineer, investigated this phenomenon in 1820. He came to the conclusion that current flowing through a wire produces a magnetic field around the wire. This field is circular. If an electric current in a wire produces a magnetic field, would a bundle of wires carrying the same current produce a much stronger field? This proved to be so. The number of wires can be increased by putting them in the form of a coil.

Section B

Ⅰ. 1. B 2. C 3. A 4. D 5. B

Ⅱ. 1. identical

2. a cylindrical core of iron laminations

3. the pole shoes, the pole faces

4. greatest, zero

5. shifting the brushes

Ⅲ. 1. The poles are mounted on the frame or yoke for two reasons: 1) it is a portion of the magnetic circuit, and 2) it can act as a mechanical support for the machine as a whole.

2. Because by doing this, the magnetic reluctance can be made relatively small.

3. For a self-excited generator, the exciting current is supplied on its own; while for a separately excited generator, the exciting current is supplied from an external

source.

4. When a commutator segment passes under a brush, the current in the armature conductor connected with it must reverse from a given value in one direction to the same value in the opposite direction. This is called commutation.

5. There are two ways of solving the sparking problem. First is to induce a voltage equal and opposite to that caused by the change in current. But a more satisfactory way is to use an interpole or commutating pole whose winding carries the same current as the armature.

Ⅳ.

English	Chinese
lamination	迭片结构
interpole	换向极
brush-holder	电刷支架
exciting current	励磁电流
in series	串联
spring pressure	弹力
air gap	气隙
mutual induction	互感

Ⅴ. 1. d 2. a 3. e 4. c 5. b 6. f

Ⅵ. 1. 当导体以切割磁力线的方式穿过磁场时,导体内部会产生电动势。

2. 电流在磁场内沿导体流动时,会产生有移动导体倾向的机械能。

3. 直流发电机包括一场结构——环形周边空间的一系列交错的南北磁极。

4. 换向器由大量铜片构成,使用屏蔽式云母片相互绝缘,且同时绝缘于支架及钳式法兰。

5. 电刷通常由石墨化的碳制成,镶嵌于金属电刷支架中,而且靠弹力顶住换向器。

Section C

Ⅰ. 1. T 2. F 3. T 4. F

Ⅱ. 1. electric circuits, magnetic circuit

2. an a.c. supply, the load

3. proportional

4. hysteresis loss, eddy-current loss

Ⅲ. 1. Transformers are used not only in substations to step-up or step-down voltages, but also frequently used in most luminous discharge tubes and bell systems in the home, as well as in radio television equipments and telephone systems.

263

2. A no-load operation means that the primary winding is connected to an a.c. supply, but the secondary winding is open.

3. In transformers, special grades of steel alloyed with silicon are usually used to give low eddy-current and hysteresis losses.

4. The five types of transformers are: 1) dry natural cooled; 2) dry forced-air cooled; 3) oil-immersed self-cooled; 4) oil-immersed forced-air cooled; and 5) oil-immersed water cooled.

Ⅳ. When a car is started, currents in an electric starter motor generate magnetic fields that rotate the motor shaft and drive engine pistons to compress an explosive mixture of gasoline and air. The ignition spark comes from an electric discharge, which makes up a momentary current flow. The phenomenon seems complex, but it actually derives from some fundamental laws of electromagnetism. One of the most important of these is Coulomb's Law, which describes the electric force between charged objects.

Ⅴ. 现代生活的每一天都充斥着电磁现象。打开电灯,电流会流过灯泡中的细灯丝。电流将灯丝加热至足够高的温度从而使其发光并照亮四周。电子钟及电子连接设备将诸如此类的简单装置连入复杂的系统,如时间与车流速度同步的交通信号灯等。收音机和电视机以电磁波的形式接收信息,电磁波以光速在空间中传输。

Unit 5

Section A

Ⅰ. 1. home robotics

2. Coded signals

3. Home energy management systems

4. architecture and engineering

Ⅱ. 1. Because its computer systems can monitor many aspects of daily living.

2. *Xanadu* 2.0.

3. First are sensors, actuators, and appliances that obey commands and provide status information. Second are protocols and tools that enable all of these devices, regardless of vendor, to communicate with each other.

4. Team Ontario described their engineering as an "integrated mechanical system" controlled by mobile devices, and it was called ECHO.

Ⅲ.

English	Chinese
smart home	智能家居
remote-controlled device	遥控设备

续上表

English	Chinese
domotics	居家机器人
communication hub	通讯集线器
science fiction	科幻小说
prototype	原型
ventilation system	通风系统
lighting	照明

Ⅳ. 1. 智能家居是一套拥有极其先进的自动化系统来控制和监控屋内任何功能设施的居家系统，能够控制照明、温度、多媒体、安全、门窗开闭、空气质量，或执行居家者其他必要或舒适的任务。

2. 1999年的一部迪士尼电影《智慧之家》讲述了一个美国家庭的滑稽行为，这个家庭赢得了配有机器人女佣的"未来之家"，而这个机器人女佣给这个家庭带来了巨大麻烦。

3. 家庭自动化领域涵盖了智能家居技术的所有阶段，包括使用高度复杂的传感器和控制设备来监控和自动调节温度、照明、安全系统和许多其他功能。

Section B

Ⅰ. 1. B 2. A 3. B 4. A 5. D

Ⅱ. 1. internal combustion

2. high energy density

3. electric motors, position

4. the health of the batteries

5. 1996，General Motors

Ⅲ. 1. Electric motors in EVs are responsible for generating power. They are also responsible for determining the total power output and the performance of an electric car.

2. Inverters in electric cars are used to convert the DC supply coming from the battery to an AC and then transfer it to the motor.

3. Firstly, batteries in electric cars store energy. When the car is turned on, the inverter converts this energy into AC form and then transfers it to the motors. A single-speed transmission is used to control the amount being transferred to the wheels. This way the power is sent to the wheels and the electric car moves forward.

4. Governments of many countries are showing their support for electric vehicles. Some prominent countries have already announced their deadlines for completely shifting

towards electric vehicles, while other countries are also planning to shift towards electric mobility.

Ⅳ.

English	Chinese
battery pack	电池组
fuel tank	燃油箱
gearbox	变速箱
transmission system	传动系统
RWD	后轮驱动
ICE	内燃机
regenerative braking	再生制动
electric vehicle	电动汽车

Ⅴ. 1. b 2. f 3. a 4. e 5. d 6. c

Ⅵ. 电动汽车最早发明于18世纪,但其从未成为主流汽车。在近十年以前,电动汽车技术一直未曾受到重视。第一辆现代电动汽车由通用汽车公司于1996年生产,此后历经近16年才发掘出其真正潜力。2012年,特斯拉在业内首次亮相,改变了汽车行业的整个局面。特斯拉进军汽车领域之后,引领了汽车行业的变革,现今市场上已有多种类型的电动汽车。其中,有几款电动汽车性能极佳,为其竞争对手内燃机汽车带来了压力。

Section C

Ⅰ. 1. F 2. T 3. T 4. F 5. T 6. F

Ⅱ. 1. Quantum levitation is to use the properties of quantum physics to levitate an object (specifically a superconductor) over a magnetic source (specifically a quantum levitation track designed for this purpose).

2. The Meissner effect dictates that a superconductor in a magnetic field will always expel the magnetic field inside of it, and thus bend the magnetic field around it.

3. A superconductor is a material in which electrons flow with no resistance.

4. Because when a superconductor tries to avoid any contact with the magnetic field, it will levitate. But the levitation isn't stable, and the levitating object won't normally stay in place. In order to make the levitation more stable, quantum locking is introduced.

5. The working principle of a frictionless bearing is that the bearing would be able to rotate, but it would be suspended without direct physical contact with the surrounding housing so that there wouldn't be any friction.

Ⅲ.

English	Chinese
equilibrium	平衡
discrete quantities	离散量
electromagnetic levitation train	磁悬浮列车
diamagnetism	抗磁性
grain boundary	晶界
counteract	抵消,中和
vortex	涡旋
quantum levitation	量子悬浮

Ⅳ. 沿着铁轨两侧安放强力电磁铁产生磁浮力。当列车运行时,同性磁极始终相对而立,列车悬浮在轨道上方几英寸处,这就是列车和铁轨之间相互排斥的磁场所产生的所谓的"磁悬浮"。由于车轮和铁轨之间没有直接接触,摩擦力为零,列车以每小时300英里的速度前进,也没有任何碰撞声。

Unit 6

Section A

Ⅰ. 1. A mechanism is a device designed to transform input forces and movement into a desired set of output forces and movement.

2. Mechanisms generally consist of moving components such as gears and gear trains, belt and chain drives, cam and follower mechanisms, and linkages as well as friction devices such as brakes and clutches, and structural components such as the frame, fasteners, bearings, springs, lubricants, and seals, as well as a variety of specialized machine elements such as splines, pins, and keys.

3. Machines have some or all of the four principal applications: 1) transform energy; 2) transfer energy; 3) multiply and/or change direction of force; and 4) multiply speed.

Ⅱ.

English	Chinese
assembly	装配,组装
lubricant	润滑油
spline	花键
gear train	齿轮传动链
machine element	机械零件
pulley	滑轮

267

续上表

English	Chinese
winding mechanism	绕线机制
clutch	离合器
linkage	连杆机构

Section B

Ⅰ. 1. T 2. F 3. T 4. T 5. F 6. T 7. T 8. F 9. F 10. T

Ⅱ. 1. rotational torque

2. gear trains

3. pinion

4. Miter gears

5. positioning, lifting

6. the sun gear, the planet gears, the ring gear

7. 4:1

8. parallel, crossed

Ⅲ. 1. There are different types of gears. For example, spur gears, helical gears, bevel gears, worm gears, rack-and-pinion gears, and planetary gears. Rack-and-pinion gears are frequently used in steering mechanism.

2. Compared with spur gears, helical gears are quieter and stronger, but they are more difficult and expensive to make.

3. One distinctive feature of the worm gear is its mechanical advantage, and the other is that in the majority of the time, they don't back drive.

4. Planetary gears are mostly used in places where a significant mechanical advantage is needed but there isn't much space. For examples, they are used in an electric screwdriver or an electric drill.

5. Chain drives are distinguished by good efficiency and no slip.

Ⅳ. A. rack-and-pinion gears

B. worm gears

C. helical gears

D. bevel gears

Ⅴ.

English	Chinese
electric drill	电钻
screwdriver	螺丝刀

续上表

English	Chinese
lever	杠杆
blender	搅拌机
linear motion	线性运动
reciprocating motion	往复运动
worm gear	蜗轮
miter gear	等径伞齿轮
velocity ratio	速率比
torque	转矩,扭矩

Ⅵ. 1. f 2. c 3. a 4. d 5. b 6. e

Ⅶ. 1. 齿轮是以不同速度、扭矩或不同方向将旋转力传递给其他齿轮的机械部件。根据其结构和设置,直径不等的齿轮可以通过组合产生正传动比,以使后一个齿轮获得不同于前一个齿轮的转速和扭矩。

2. 斜齿轮具有一定的优势。例如,在连接平行轴时,斜齿轮比齿数相同的正齿轮具有更高的承载能力。

3. 链条传动与开口皮带传动类似,适用于中心距大于齿轮传动的平行轴。用链条将驱动轴与从动轴上的链轮连接在一起,轴与轴之间输入与输出的转动方向可以是相同的,也可以是相反的。

Section C

Ⅰ. 1. C 2. B 3. A 4. B 5. C 6. B 7. D

Ⅱ. 1. structural components, mechanisms, control components

2. the surface finishes, sliding velocities

3. ball bearings, roller bearings, needle bearings

4. Belleville washers / springs

5. angular misalignment, parallel offset

6. engaged

7. the engine, the gearbox

8. magnetic force

Ⅲ. 1. Friction results in loss of power, the generation of heat, and increased wear of mating surfaces.

2. The chief characteristics of springs are the ability to tolerate large deformations without failure and to recover its initial size and shape when loads are removed.

3. The flange coupling is the most common coupling. It has the advantage of simplicity and low cost, but the connected shafts must be accurately aligned to avoid severe

bending stress and excessive wear in the bearings.

4. Clutches are connected to two rotating shafts. One shaft is attached to a motor or other power unit (the driving member) while the other shaft (the driven member) provides output power for work to be done.

5. The clutch of a belt-driven engine cooling fan is heat-activated. When the temperature is low, the spring winds and closes the valve, which allows the fan to spin at about 20% to 30% of the shaft speed. As the temperature of the spring rises, it unwinds and opens the valve, allowing fluid past the valve which allows the fan to spin at about 60% to 90% of the shaft speed.

Ⅳ.

English	Chinese
radial load	径向载荷
surface finish	表面抛光
vibration damping	减振
bending stress	弯曲应力
powertrain	动力系统
separator	轴承座
deformation	变形
gearbox	变速箱
cantilever spring	悬臂弹簧
default state	缺省状态
shearing stress	剪应力
manual transmission	手动变速箱

Ⅴ. 1. specifically   2. misalignment   3. flexibility   4. cushion
   5. diameter   6. disengage   7. hydraulic

Ⅵ. 1. There are various types of bearings, such as sliding bearings, ball bearings, roller bearings, and the like. We use different types of bearings for different purposes. Although the responsibility for the basic design of ball and roller bearings rests with the bearing manufacturer, the machine designer must form a correct appreciation of the duty to be performed by the bearing. He should not only concern the selection of bearings, but also the conditions for correct installation.

2. Transmissions, shafts, bearings, and other components are typical examples of machine elements that are used in a wide variety of different applications. Transmissions are designed to reduce or increase a speed-torque to achieve a suitable

output power. Shafts, bearings, and other components are also employed in all kinds of machinery and mechanical equipments.

Section D

Ⅰ. 1. T 2. F 3. F 4. F 5. F 6. T

Ⅱ. 1. The relationship between joints, links, and a linkage can briefly be summarized as follows: the rigid bodies are called links, and they can be connected by joints; a linkage is an assemblage of links connected by either revolute joints or prismatic joints.

2. The four types of joints and their respective examples are as follows:

　　1) Revolute joint. A good example is the hinge used to attach a door to the frame.

　　2) Prismatic joint. The connection between a piston and a cylinder in an internal combustion engine or a compressor is via a prismatic joint.

　　3) Spherical joint. A good example is the shoulder and hip joints of the human body.

　　4) Helical joint. A good example is the relative motion between a bolt and a nut.

Ⅲ. 1. d 2. b 3. f 4. e 5. c 6. a

Ⅳ. 1. 连杆是刚体或构件连接在一起而形成的运动链。如果不考虑机器构件应变所造成的细微偏差,刚性连杆(有时简称连杆)是机制研究的理想状态。

2. 连杆机构可以被定义为固体结构或连杆的组合,其中每个连杆通过销接或滑动关节与至少两个其他连杆相连。为了满足这一定义,连杆机构必须形成一条无限的或封闭的链或一系列封闭的链。

3. 四连杆机构是一种适用于自行车的机械连杆机构。常规全悬挂自行车的后轮在极小的弧度内运动,这意味着上坡时会损耗更多动力。装配有四连杆机构的自行车,其车轮运动弧度很大,几乎是垂直运动,这样可以将功耗降低30%。

Unit 7

Section A

Ⅰ. 1. B 2. C 3. A 4. D 5. C 6. C

Ⅱ. 1. c 2. d 3. f 4. b 5. e 6. a

Ⅲ. 1. Relative density is the density of a material compared with that of the water at 4℃.

2. Fusibility is the ease with which materials melt. Fire bricks have low fusibility because they melt at very high temperatures and are not easy to melt.

3. Creep is the gradual extension of a material over a long period of time while the applied load is kept constant. The creep rate increases if the temperature rises, but becomes slower if the temperature drops.

4. The distinct difference between malleability and ductility lies in the property of the

applied load. For malleability the applied load is the compression load, while for ductility the tension load.

5. 1) If a material is within its elastic range, when the applied load is removed, it will return to its original shape and size;

2) If a material is beyond its elastic range but within its plastic range, when the applied load is removed, it will deform permanently;

3) If a material is beyond its plastic range, it will break or rupture.

Ⅳ.

English	Chinese
malleability	展延性
rolling	轧制
stiffness	刚度
cast iron	铸铁
corrosion resistance	耐腐蚀性
flashing	防水板
fatigue strength	疲劳强度
furnace lining	炉衬
tube drawing	管材拉拔

Ⅴ. 1. 良好的磁性导体磁阻较低,例如铁磁材料,它们是由铁、钢和相关合金元素(如钴和镍)制成,并据此被称为铁磁材料。

2. 蠕变是指在外加载荷不变的情况下,随着时间的延长,材料逐渐延展。

3. 展延性是指材料承受压缩变形而不破裂的能力,或者是指展延性材料在断裂发生之前,能够在压缩载荷下承受一定量的塑性变形。

4. 部分诸如铅之类的金属在室温下具有良好的塑性范围,可用于广泛加工。这一特性有利于水管工在建筑工地上敲打铅防水板进行成型操作。

Section B

Ⅰ. 1. F 2. F 3. T 4. T 5. F 6. T

Ⅱ. 1. iron

2. iron, carbon

3. tools, dies

4. hardness penetration

5. the hot working, the impact properties

6. warping, scaling

7. bar, rod, structural

8. white iron, gray iron, malleable iron

Ⅲ. 1. Engineering materials can be divided into metals and nonmetals. According to whether they contain iron or not, metals can be subdivided into ferrous metals and nonferrous metals. According to whether they are associated with natural substances, nonmetals can be subdivided into organic materials and inorganic materials.

2. Because the amount of carbon present largely determines the maximum hardness obtainable. The higher the carbon content, the higher the tensile strength and the greater the hardness to which the steel may be heat-treated.

3. Alloy steels are defined as steels containing very small quantities of elements other than carbon, phosphorus, sulphur, and silicon. A car is made of 100 different kinds of alloy steels.

Ⅳ.

English	Chinese
annealing	退火
trace	痕量
finish	光洁度
tungsten	钨
hardness penetration	淬硬深度
rolled steel	轧钢
drill rod	钻杆
ferrous metal	黑色金属
alloying agent	合金添加剂

Ⅴ. 1. c 2. f 3. a 4. b 5. d 6. e

Ⅵ. Characteristically, metals are opaque, ductile, and good conductors of heat and electricity. Metals are divided into ferrous metals and non-ferrous metals. The former contain iron and the latter do not. The addition of certain elements into steel can improve its properties. For example, chromium may be included to resist corrosion and tungsten to increase hardness. Aluminum, copper, alloys, and brass are common nonferrous metals.

Section C

Ⅰ. 1. D 2. D 3. C 4. A 5. D 6. B

Ⅱ. 1. Ease of fabrication

2. thermosetting, thermoplastic

3. they do not cure or set

4. diamond, graphite

5. electrically, thermally

6. requiring extensive machining

7. directional-strength

8. fusing clays, earthy

Ⅲ. 1. Nonmetallic materials can be divided into organic and inorganic materials according to whether they are associated with natural substances or not. Organic materials can be subdivided into polymers and others, and the former of which include thermosetting plastics(thermostes), thermoplastics, and elastomers. Inorganic materials can be divided into ceramics, glasses, and others.

2. Rubbers are natural materials, and synthetic rubbers are polymers which have been synthesized to reproduce consistently the best properties of natural rubbers.

3. The production of carbon and graphite components involves two processes: (1) molding or extrusion followed by oven baking; (2) machining. The first process is economically preferred to mass production.

Ⅳ.

Information	Thermosets	Thermoplastics
when heated	They will be cured, set, or hardened.	They will be in a flowable state.
characteristics	The curing is an irreversible chemical reaction.	They can be remelted and rehardened by cooling many times, but is limited by thermal aging.
in fabrication	They can be removed from the mold at the fabrication temperature.	They must be cooled in the mold to prevent distortion.

Ⅴ. 1. metallurgical

2. malleable

3. susceptible

4. recrystallization

5. irreversible

Ⅵ. 1. "弹性材料"这一术语包括弹性或橡胶状聚合物的全部领域。有时随机称为橡胶、合成橡胶或弹性材料。但是,更恰当地说,橡胶是天然材料,而合成橡胶是合成的聚合物,用以复制天然橡胶的最佳性能。由于存在大量的橡胶状聚合物,"弹性材料"这一广义的术语是最合适和最常用的。

2. 在需要广泛加工的应用中,石墨比碳更具有优势。石墨非常容易加工,其公差可以达到金属粗加工的程度。在只需要切割加工或其他最少加工时,推荐使用碳素构件。

Unit 8

Section A

Ⅰ. 1. F 2. F 3. F 4. F 5. F 6. T 7. T

Ⅱ. 1. e 2. a 3. b 4. c 5. d

Ⅲ. 1. Heat treatment is a process utilized to change certain characteristics of metals and alloys by either heating or cooling in order to make them more suitable for a particular kind of application.

2. The critical temperature for steel is the point at which it has the most desirable characteristics. The temperature of 1400°F-1600°F is ideal for steel to make for a hard, strong material if it is cooled quickly.

3. The most obvious drawback of hardening is the internal strain produced in the process of hardening, which could cause the metal to crack. In order to solve the problem, tempering process is followed.

4. Three different kinds of annealing processes are stress-relief annealing, spheroidized annealing, and full annealing.

Ⅳ. Low carbon steels do not become hard when subjected to such a heat treatment because of the small amount of carbon contained. If it is necessary to obtain a hard surface on a part made of steel, case hardening operation must be carried out. One of the methods of case hardening is cyaniding, which is done by keeping the work in a molten bath of sodium cyanide from 5 to 30 minutes. Having been subjected to such a treatment, the work is then quenched in water or oil, and a very hard case of 0. 254 to 0. 381 mm thick is formed.

Section B

Ⅰ. 1. cold forming, hot forming

2. plastic deformation, brittle

3. rolled state, stock

4. open

5. the closed die forging

6. several passes

Ⅱ. 1. b 2. e 3. d 4. c 5. a

Ⅲ. 1. HF 2. HF 3. CF 4. CF 5. HF 6. CF 7. HF

Ⅳ. 1. Forming can be defined as a process in which the desired size and shape are obtained through the plastic deformation of a material. The typical forming processes are

rolling, forging, extrusion, drawing, and deep drawing.
2. The two broad categories of forming processes are cold forming and hot forming. If the working temperature is higher than the recrystallization temperature of the material, then the process is called hot forming. Otherwise the process is termed as cold forming.
3. Because hot working results in shrinkage and loss of surface metal due to scaling. Moreover, surface finish is poor due to oxide formation and scaling.

V.

English	Chinese
closed die forging	闭式模锻
intermediate annealing	中间退火
flash	飞边
coefficient of friction	摩擦系数
cast ingot	铸锭
blank holder	压边圈,压料板
drawing	拉拔
extrusion	挤压
strain hardening	应变硬化

Ⅵ. 1. 此外,一部分输入的能量在通过应变硬化提高产品的强度上得到了卓有成效的利用。
2. 在闭式模锻中,通过在两个成型模和闭式模之间挤压工件来获得所需的形状。
3. 大量的线材、棒材、管材和其他型材是通过拉拔工艺生产的,拉拔工艺基本上是一种冷加工工艺。
4. 金属薄板被放在模具上,通过坯料压板来避免生产中的产品瑕疵。
5. 然而,冷加工除了其他局限性以外(如难以加工高强度和脆性材料以及大的产品尺寸),该工艺还有一个不良特性,即无法防止防腐蚀性能明显减小。

Section C

Ⅰ. 1. C   2. D   3. A   4. C   5. A   6. C
Ⅱ. 1. d   2. a   3. g   4. b   5. f   6. e   7. h   8. c
Ⅲ. 1. Casting is the introduction of molten metal into a cavity or mold where, upon solidification, it becomes an object whose shape is determined by mold configuration. The advantages of casting are as follows: it is adaptable to intricate shapes, to extremely large pieces, and to mass production; it can provide parts with uniform physical and mechanical properties throughout; and it also has economic advantage.

2. Because in semi-continuous cast ingots, the process will cease after a certain length of time, but in continuous cast ingots, the solidified ingot is continually sheared or cut into lengths and removed during casting. Thus the process is continuous, the solidified bar or strip being removed as rapidly as it is being cast.

3. Carbon soot or refractory slurry is coated on the surface of the mold cavity in order to increase mold life, to make ejection of the casting easier, to serve as heat barriers, and to control the rate of cooling of the casting.

4. Rheocasting is the casting of a mixture of solid and liquid. In this process, the alloy to be cast is melted and then allowed to cool until it is about 50% solid and 50% liquid. A major advantage of rheocasting is the much reduced die erosion due to the lower casting temperatures.

Ⅳ.

English	Chinese
casting to shape	铸型
centrifuging	离心法
solidification	凝固
symmetry	对称
internal cavity	内腔
parting line	分型线
bulk density	体积密度,松装密度
root mean square	均方根
static cast ingot	静态铸锭

Ⅴ. 1. The traditional casting is still an important process adaptable to intricate shapes, to extremely large parts and very small pieces.

2. Investment casting is relatively applicable to the manufacture of special cavities and cores.

3. Theoretically speaking, the parts manufactured by the identical mold are interchangeable parts.

4. Permanent-mold casting techniques can be used to replicate intricate and complex patterns ranging from small to large pieces.

Unit 9

Section A

Ⅰ. 1. B 2. B 3. D 4. A 5. D

Ⅱ. 1. lean production processes
2. Quality control
3. turnkey solution
4. e-commerce sales
5. giving verbal commands

Ⅲ. 1. The Covid-19 pandemic created a demand for automation in various business sectors, including those outside industrial manufacturing.
2. The examples of sanitation tasks done by cleanroom robots in manufacturing are preparing syringes for coating, placing pills into bottles and capping them, and handling food packaging.
3. Companies using voice-picking AMRs report having immediate gain in productivity and savings with annual labor costs.

Ⅳ.

English	Chinese
quality inspection	质量检验
lean production	精益生产
collaborative robot	协作机器人
workflow	工作流程
supply chain	供应链
social distancing	社交距离
spectral imaging	光谱成像
cutting-edge technology	尖端技术

Ⅴ. 1. 由于许多产品面临供应短缺，制造业、生命科学、食品加工、零售和配送等行业都将自动化视为解决方案。这些需求促使了自动化在新领域的运用和在常规领域的提升。
2. 视觉质检技术为产品提供了额外的保障，确保所有离开生产线到达消费者手中的产品都是最高质量的。
3. 公司正在以战略方式实施协作技术，以克服雇佣工人不足、社交距离限制以及对精益生产流程的迫切需求。
4. 机器人已经在供应链所有环节中普及，但随着电子商务物流量的增加，必须在仓库和配送中心使用移动机器人。这些机器人可以在没有人工干预的情况下自主改变路线，并且可以与工人协作执行任务。

Section B

Ⅰ. 1. scanning probe microscopes

2. unusual properties
3. kill bacteria
4. computers
5. gold nano shells

Ⅱ. 1. Nanotechnology is the understanding and control of matter at dimensions between approximately 1 and 100 nanometers, where unique phenomena enable novel applications. Encompassing nanoscale science, engineering, and technology, nanotechnology involves imaging, measuring, modeling, and manipulating matter at this small scale.

2. The three basic concepts of the NNI definition are: 1) Nanotechnology is very, very small; 2) At the nanoscale, materials may behave in different and unexpected ways; and 3) Researchers want to harness these different and unexpected behaviors to make new technologies.

3. Because they want to communicate Tata Nano's use of high technology and small size to its potential buyers.

Ⅲ.

English	Chinese
sun block	防晒霜
light microscope	光学显微镜
circuitry	电路系统
gel	凝胶
nanometer	纳米
clinical trial	临床试验
antimicrobial	抗菌剂
inventory	清单,存货

Ⅳ. 1. 新闻报道已经预示纳米技术将成为下一场科学革命——比如研发出更快的计算机、治疗癌症和解决能源危机等。

2. 通过运用这些新属性,不同学科的研究人员希望创造出新事物,从诸如抗菌袜子、轻质网球拍等日常用品到最先进的太阳能电池、更小更快的计算机或是选择性治疗疾病的医疗手段。

3. 纳米技术被称为"下一件大事",它被誉为癌症治疗、能源自主、电子设备改进以及为第三世界国家带来清洁水的关键。

Section C

Ⅰ. 1. T 2. T 3. F 4. T 5. F 6. T

II. 1. 3D printers build objects using a process known as additive manufacturing. Material is put down in layers; each layer adds to the previous layer and in turn becomes a base for the next layer.

2. Some 3D printers incorporate USB ports to read files from USB drives; other 3D printers interface to external computers, which may be running 3D print monitor and control applications.

3. Today 3D printers will face limits in what can be manufactured as well as uncertainty within the legal framework surrounding print files and printed objects.

III. 1. f 2. a 3. e 4. c 5. b 6. d

IV. 3D printing is a kind of rapid prototyping technology, also known as additive manufacturing. Based on digital model files, 3D printing technology uses adhesive materials such as powder metals or plastics to build objects through printing layer by layer. 3D printing is usually realized by using digital technology printers. It is often used in mold manufacturing, industrial design and other fields to produce models, and then gradually used in product manufacturing. There are already parts printed by this technology. 3D printing technology has wide applications in jewelry, footwear, industrial design, architecture, engineering and construction, automobile, aerospace, dental and medical industry, education, civil engineering, firearms, and other fields.

Unit 10

Section A

I. 1. T 2. F 3. T 4. F 5. T

II. 1. Optimists foresee unlimited energy through harnessing the power of the atom, the discovery of new and unlimited food sources, and the dawn of an age in which human drudgery is replaced by technological advances.

2. With electron accelerators, electrical and electronic scientists and engineers will probe the mysteries of the atomic nucleus, and with radio telescopes they will study signals from remote regions of outer space.

3. Inertial guidance systems are capable of guiding rockets and interplanetary spaceships by using devices which will detect changes in speed and direction and make necessary adjustments automatically.

4. Fueling aircraft and spacecraft by laser beam would greatly reduce the weight of aircraft and spacecraft and thereby increase the probability of hypersonic travel.

Ⅲ.

English	Chinese
depletion	耗尽
cybernetics	控制论
electron accelerator	电子加速器
variables	变量
interplanetary spaceship	星际飞船
radio telescope	射电望远镜
retina	视网膜
pulsar	脉冲星
transducer	换能器
inertial guidance system	惯性制导系统

Ⅳ. 1. geometric   2. inexhaustible   3. scarcity   4. envelop
   5. detached   6. hypersonic

Ⅴ. 1. 悲观主义者预测的是毁灭。他们着眼于日益加剧的大气、水和土地污染,原材料枯竭,一些现在常见能源耗尽,以及人口膨胀等问题。
   2. 假如有一个良好的经济和政治环境,这将对科学技术的发展起到巨大的促进作用。
   3. 激光已经被用于连接活体组织,如脱落的视网膜。对于那些不宜用手术刀进行的复杂而精细的外科手术,激光的使用也将变得司空见惯。
   4. 他们将改进天线(用于接收电磁波的扇形接收天线)、过滤器(用于过滤电波或电流的装置)、转换器(用于将某种能转换成另一种能的装置)和继电器(在电路中用来接通和断开开关的装置)。这些都是电子工业中的基本元件,是支撑我们经济的重要组成部分。

Section B

Ⅰ. 1. the laws of mechanics
   2. 10 to the power of -9
   3. carbon footprint, eco-friendly and economical
   4. mass dampers
   5. potential weapons of mass destruction

Ⅱ. 1. The specific applications of the future of mechanical engineering are: 1) nano-engineering; 2) biomechanics; 3) automobiles and aviation; 4) buildings of the future and urban designing; and 5) robotics.
   2. "Bionic" is the term used to refer to the artificial material or object that mimics the action done by a part of the human body. For example, a bionic arm mimics the actions of a human arm; the bionic leg mimics the human leg.

III.

English	Chinese
bionics	仿生学
inclined plane	斜面
mass damper	质量阻尼器
biomechanics	生物力学
carbon footprint	碳足迹
aviation	航空
ASIMO	阿西莫机器人
NASA	美国国家航空航天局

IV. 1. boon 2. pivotal 3. mimic 4. stellar
  5. quintessential 6. tremor 7. microscopic

V. ASIMO, the world's first humanoid robot with the ability of human bipedal walking, is the product of countless scientific and technological research invested by Honda. It has won the love of many people with its clumsy but lovely shape, and many of its humanoid functions have continuously surprised people. It seems that the plot in science fiction movies is becoming a reality step by step. With a height of 1.3m and a weight of 48kg, ASIMO walks 0-9km per hour. When early robots walked in a straight line, they had to stop before a sudden turn, which made them look clumsy. ASIMO, however, is more flexible. It can predict the next movement in real time and change the center of gravity in advance. Therefore, ASIMO can walk freely and perform various seemingly complex actions such as zigzag walking, stepping down the stairs, bending down, etc. In addition, ASIMO can shake hands, wave hands, and even dance to music.